Out of Water

Out of Water

From Abundance to Scarcity and How to Solve the World's Water Problems

Colin Chartres and Samyuktha Varma

Vice President, Publisher: Tim Moore
Associate Publisher and Director of Marketing: Amy Neidlinger
Executive Editor: Jim Boyd
Editorial Assistant: Pamela Boland
Development Editor: Russ Hall
Operations Manager: Gina Kanouse
Senior Marketing Manager: Julie Phifer
Publicity Manager: Laura Czaja
Assistant Marketing Manager: Megan Colvin
Cover Designer: Chuti Prasertsith
Managing Editor: Kristy Hart
Project Editor: Anne Goebel
Copy Editor: Language Logistics, LLC
Proofreader: Kathy Ruiz
Senior Indexer: Cheryl Lenser
Senior Compositor: Gloria Schurick
Manufacturing Buyer: Dan Uhrig

Publishing as FT Press
Upper Saddle River, New Jersey 07458
FT Press offers excellent discounts on this book when ordered in quantity for bulk purchases or special sales. For more information, please contact U.S. Corporate and Government Sales, 1-800-382-3419, corpsales@pearsontechgroup.com. For sales outside the U.S., please contact International Sales at international@pearson.com.

Printed in the United States of America
First Printing July 2010
Pearson Education LTD.
Pearson Education Australia PTY, Limited.
Pearson Education Singapore, Pte. Ltd.
Pearson Education North Asia, Ltd.
Pearson Education Canada, Ltd.
Pearson Educación de Mexico, S.A. de C.V.
Pearson Education—Japan
Pearson Education Malaysia, Pte. Ltd.
Library of Congress Cataloging-in-Publication Data
Chartres, Colin John, 1951-
Out of water : from abundance to scarcity and how to solve the world's water problems / Colin Chartres,
Samyuktha Varma. — 1st ed.
p. cm.
ISBN 978-0-13-136726-5 (hardcover : alk. paper) 1. Water resources development. 2. Sustainable development. 3. Climatic changes—Environmental aspects. I. Varma, Samyuktha, 1981- II. Title.
HD1691.C493 2010
333.91—dc22
2010014840
ISBN-10: 0-13-136726-9
ISBN-13: 978-0-13-136726-5

This book is dedicated to Margaret, Andrew, and Christopher Chartres for their long standing support of Colin's journey through the world of water, which often takes him far from home.

Contents

Acknowledgments

This book draws from over 25 years of research from the International Water Management Institute and especially the findings of an innovative five-year process it led involving over 700 scientists published in *Water for Food, Water for Life: A Comprehensive Assessment of Water Management in Agriculture*. The book also credits longstanding partnerships with committed government agencies, water planners, and national agriculture and water research institutes across the developed and developing world that have enabled many of the lessons for good management and practice to come to light. In addition, stimulating discussions with colleagues from institutes including the FAO, other CGIAR Centers, the RAMSAR Convention, UNESCO, the Global Water Partnership, UN-WATER, CSIRO, the Australian National Water Commission, the Stockholm International Water Institute, McKinsey and Company, and WWF are gratefully acknowledged. Many of these people have worked hard to put water issues on the global agenda and acted with the foresight to demonstrate that it takes many types of solutions and many more types of people to solve the water crisis.

Particular thanks go to Dr. David Molden (IWMI), Dr. Mark Giordano (IWMI), and Will Fargher (Australian National Water Commission) for critically reviewing parts of the book. Additionally, special thanks go to IWMI's Dr. Tushaar Shah for assistance with the water economics sections of Chapter 9. The authors would also like to thank IWMI's Nishadi Eriyagama and Joanna Kane-Potaka for their help. We have also tried hard to faithfully acknowledge original source materials in the endnotes. The section on U.S. water rights in Chapter 9, "Water Rights and Water Costs," relies heavily on material posted on the U.S. National Science and Technology Center of the Bureau of Land Management's website. Given the legal nature of this material, we changed relatively little to ensure that correct meaning was not lost. We thank the International Food Policy Research Institute (IFPRI); UNEP-GRID Arendal; Department of the Environment, Water, Heritage and the Arts of the Government of Australia; Water Corporation, Western Australia; and the Intergovernmental Panel on Climate Change (IPCC); Data 360, for granting permission to reproduce some of the graphs and maps in this book.

Samyuktha would like to thank her parents for making sure her many interests were always encouraged.

About the Authors

Colin Chartres is the Director General of the International Water Management Institute, which is headquartered in Colombo, Sri Lanka, and operates across Asia and Africa. He has over 30 years experience working in natural resources management in the private sector, academia, research institutions, and government. A keen traveler and trekker, he is acutely aware of the increasing impact that humans are having on the environment and that science-based solutions, placed in the right socio-economic context, can alleviate some of these impacts and improve the livelihoods of poor people. Colin began his career as a geographer and soil scientist and has BSc and PhD degrees from the Universities of Bristol and Reading in the UK and before moving to Sri Lanka, lived in Australia for 25 years. More recently, he has been involved in water resources management and policy issues, but overall he believes that natural resource and environmental issues need scientifically integrated solutions. He has published over 120 journal articles and book chapters.

Samyuktha Varma was born in India, educated in the United States and the UK, and now lives and works for IWMI in Sri Lanka. Her background in international relations and gender and social policy has led to her working on a range of issues from human rights advocacy to urban governance, and presently in the world of water with IWMI. As a social scientist she has explored the question of equity in water management, particularly the impact of development interventions and policies on social equity. As a writer, she has written policy and science briefs and been involved in the communication of research for development. She has written several papers and articles on intersectoral issues and governance of water.

IWMI is one of 15 international research centers supported by the network of 60 governments, private foundations, and international organizations collectively known as the Consultative Group on International Agricultural Research (CGIAR). IWMI has a staff of about 270 and offices in 12 countries across Asia and Africa. Its Vision is "Water for a food secure world." It focuses on the physical issues of water scarcity, availability, productivity, and the socio-economics of water including governance, institutions, and gender issues with particular focus on the poor and agricultural development. IWMI celebrated its 25th anniversary in 2010.

Foreword

Water is fundamental for life and the environment. The economies of all nations are dependent on access to water, and its use is indispensable across all economic sectors. However, water has for far too long been taken for granted. Climate change points at the problem, as water is the problem factor in adaptation efforts to climate change.

For the case of crude oil, we talk about the concept of "peak oil" approaching in the next few years. While we have an expectation that technology will rise to the challenge of replacing oil with more sustainable energy sources, we all know and understand that this will require billions of dollars of investment in research and development undertaken by the private sector and governments. The world is similarly approaching "peak water," but the solution may be even more difficult than in the case of oil. In many water-scarce countries, water demand is predicted to significantly outpace supply over the next few decades. This will have profound impact and none more so evident than on food production.

At the turn of the last century, approximately 850 million people were living below the poverty line. The recent global financial crisis, on top of the 2007–2008 food crisis, has seen this number rise to over 1 billion. These people do not have the money or the means to provide themselves and their families with adequate nourishment and clothing, let alone attend to basic health needs. This problem is often exacerbated by malnutrition and disease. If we are to feed the increasing numbers of people on the planet by 2050, we are going to require about twice as much water as was used in agriculture in the year 2000. The competition for water from dietary change, urbanization, and industry and the risk that climate change will adversely impact water supplies in some countries is a challenge that is unlikely to be met and threatens to push even more people into abject poverty. Additionally, in some countries, government mandated targets for biofuel production also threaten water and land resources previously used for food production.

A thesis at the heart of this book is that all farmers, but particularly those in water-scarce developing countries, have to increase their water use efficiency. However this will be impossible to achieve without increased investment. Similarly, agricultural development has repeatedly shown as a sound basis for economic growth, and it particularly benefits the vast numbers of smallholder farmers that constitute the poorest and rely primarily on agriculture for their livelihood. And typically these groups also pay implicitly the highest unit prices for drinking water, be it in time cost or because of water vendors.

The authors indicate that we must stop taking the "business as usual" approach to water and food production. They demonstrate that there are multiple and complementary solutions to increasing the productivity of water and food production. These solutions are both based in science and engineering and in economics and policy. As in the case of seeking solutions to the peak oil problem, seeking solutions for peak water will need major investment in infrastructure, policy, and governance reform and in changing the institutions that manage water and in building the capacity of poor farmers, many of whom are women. These changes will require that we put the needs of the poorest and the most vulnerable at the heart of the solutions. Without this any sustainable change to improve both the management of water as well as the wellbeing of smallholders will not succeed. These solutions also cannot be thought as panaceas that will work on their own. In order to improve the way farmers grow food, we also need to assist in the way it is delivered to markets—and the capital to grow and raise yields must also be available and function. The solutions in this book also highlight how important these links are and how inseparable the management of water is from so many other things. We therefore need to further explore the potentials for water use efficiency along the whole water value chain.

Combating food security problems and ensuring that we use water more wisely, more productively, and less profligately will be one of the defining challenges facing policy makers and managers of natural resources and of agriculture this century. There are solutions, and while implementing these will not be easy, they are achievable, as pointed out in this book.

Dr. Joachim von Braun
Director, Center for Development Research (ZEF)
Professor for Economic and Technological Change,
University of Bonn, Germany
Former Director General, International Food Policy
Research Institute

Preface

The year 2008 was one of crises: a food crisis, a fuel crisis, and a financial meltdown. While the impacts of these crises are being felt across the globe, they all have ramifications with respect to natural resources and, in particular, water. In many regions of the world, the food crisis was precipitated by drought and increasing water scarcity, which triggered a supply-demand imbalance and resulted in the consequent price hike. The fuel crisis stimulated demand for biofuel, which requires vast areas of land and volumes of water for its production, prompting many countries to grow biofuel crops to meet "green energy" targets without adequate consideration of the impact on water resources and food production. The impact of the financial crisis on investment in international aid, water delivery, and storage infrastructure such as dams, reservoirs, and piped supply systems will undoubtedly be felt by water users around the world, particularly the poor, as tightening credit limits finances available for such development. This leaves us with the critical question, do we also have a water crisis as has been suggested by some recent commentators and if so, how severe is the global water crisis?

This book is aimed at readers interested in how growing water scarcity will impact food production and the environment over the next 50 years and potential responses. It sets out to explore the key factors that control water availability, scarcity, and use and to determine what increasing demand will do to global water resources. It also looks at ways in which we can learn to use and manage our water resources better and with greater equity. As the world gears up to meet the challenges of population growth, urbanization of the developing world, climate change, a growing demand for energy, and food and environment security, we find ourselves in great need for a new approach to solving problems. This book also sets out to explore how this might be done using scientific evidence as the basis for policy decisions that will improve the management of water resources, with people and the environment at the center of the solutions. It presents some of the successes of water management initiatives, past and present, and the central role that governments and stakeholders will play in averting a crisis and planning for sustainable use of water resources in the future.

Much of the book is based around the premise that we have to manage our water resource better than we have in the past if we are going to feed a predicted world population of nine billion by the middle of the century. We cannot continue with stagnating or low growth of food and water productivity if everyone is to be fed and cities and industries are to get adequate water—along with the environment getting allocated its fair share as well. Chapter 1, "Not Another Crisis!" presents the evidence that highlights the severity of the global water crisis and whom, how, and what it will effect. In Chapter 2, "From Abundance to Scarcity in 25 Years," several case studies demonstrate how water has come to be so scarce in some regions of the world. Chapter 3, "Causes of Water Scarcity," details the "drivers" of the water crisis (population, urbanization, globalization, dietary change, and biofuel demand are explained). Given the significance of climate change, Chapter 4, "Climate Change and Water," looks in more detail at the potential impact of climate change on agriculture and water resources.

These chapters are followed by a chapter that looks at the fundamental role of agriculture (Chapter 5, "Agriculture and Water") both in terms of feeding growing populations, but also as the major user of water in most countries. Significance is given to just how developing agriculture is a key method not only of feeding everybody, but as a fundamental development and economic growth pathway. The complex relationships between water, food, and poverty are examined in Chapter 6, "Water, Food, and Poverty." Water supply and sanitation

(Chapter 7, "Integrating Water Planning and Management") are examined in the context of the Millennium Development Goals and integrated water resources management, based on the key premise that water resources and uses have to be viewed holistically and not as separate sectors of the economy if we are to overcome the looming crisis.

The latter chapters of the book turn to the thorny issues of governance and policy development (Chapter 8, "Water Governance for People and the Environment") in the water environment and the need for wholesale reform in most countries. Attention is also given to the even more emotive issues of how we allocate water and how we should be valuing, costing, and pricing water delivery to different users, taking into account the fundamental human right of access to water for drinking, cooking, and washing (Chapter 9, "Water Rights and Water Costs").

The picture painted is not rosy. However, the book concludes by looking at some potential solutions and focusing on doing business better in the future. These solutions need strong endorsement and support from all levels of society and, in particular, politicians.

In a way, the book is an idiosyncratic journey through the complex world of water based on the lifelong experience of its senior author, in natural resources science and policy and program management in government. Each step or chapter unravels another layer of complexity that has to be contended with if we are to deal effectively with increasing water security. What it intends to demonstrate to the reader is that water management is not simply a matter of understanding the hydrological cycle and laws of gravity. Rather, good water management represents a balancing of the aspirations of all parts of society whether in developed, emerging, or developing economies. While sound scientific evidence and economic theory help to underpin good solutions, at the end of the day, it is the human factor regarding acceptability of those solutions that is most important.

chapter one

Not Another Crisis!

"Water is the driver of Nature."

Leonardo da Vinci

Why Is Water Important?

Twenty years ago, it was rare for water to be in the news. Today, in some parts of the world, it is unusual not to see an article about water, or lack of it, in newspapers and influential magazines on a weekly basis. Furthermore, water shortages have occurred in areas previously considered well-endowed with rainfall including Southeast England and parts of the southeastern United States (Georgia, for instance). Authorities are now starting to realize that water supplies are not infinite, and the general public is now realizing that taking 30 minute showers and hosing down driveways might be rather profligate ways of behaving. Elsewhere, in drier regions, reduced water availability has not only caused crop failures and starvation in developing countries, but is threatening economic development in the developed and developing worlds. While climate change will undoubtedly have an increasing impact on water availability and food production over the coming decades, there are many other factors including urbanization and industrialization, people's changing diets, and biofuels production that already affects and will increasingly impact water availability. The growing twenty-first century "water gap" between supply and demand is going to have major ramifications for the entire planet. This book examines how and why it has developed, its causal factors, and where its consequences will be most severe.

In the years following the Second World War until about 1990, availability of water resources was not seen as a factor that might ultimately limit development except in the driest countries. Just as in the case of land, water resources were viewed as an almost free good there for the taking, and governments and private investors gave them little thought. Where water resources problems existed, these were generally associated with water quality. However, in the western world more stringent environmental laws, applied over the last 50 years, have seen many rivers being returned from little more than open sewers to relatively clean systems capable of supporting significant aquatic life once again.

Yet in the same time period in many countries, politicians allocated water to agricultural and other users as if it were an infinite resource. There seemed to be no realization that water could or would eventually become scarce. Many developing countries had little or no policy or regulation focused on water resources management. By the mid-1980s water resources were showing signs of stress particularly in some Middle Eastern countries and also in developed countries including Australia and the Southwestern United States. At the turn of the millennium numerous countries were facing water scarcity. Then came 2008, a year of crises, which saw a food crisis, a fuel crisis, and a financial meltdown. The 2007–2008 food crisis was primarily one of supply and demand imbalance, although world food stocks did reach alarmingly low levels. While many people don't realize it, there are direct relationships between the food crisis and water scarcity that were manifest initially through drought in various regions. The growing world population and its increasing demand for food has undoubtedly been the primary driver behind increasing demand for water. Dietary changes from cereal- and vegetable-based to more animal-based foods have also caused increasing agricultural demand for water. Water requirements for the production of a kilo of beef can be as high as 3000 gallons, as is demonstrated later in this book. Similarly, the fuel crisis pushed up the price of biofuels, thus creating competition with food production for land and water. The financial crisis may also have profound ongoing impacts on water resources availability as billions of dollars of investment are needed in many countries, particularly in sub-Saharan Africa to build water infrastructure. The extent to which the financial crisis will reduce overseas aid spending is uncertain. However, in times of financial stringency, it is much harder for all borrowers, including governments, to access funds for infrastructure development. The 2007–2008 food crisis has come and gone, but a delayed monsoon in India in 2009 again alerted authorities to the precarious relationship between food insecurity and water availability. As we move through the twenty-first century, the underlying fundamental problems of concern to water resources management include continued population growth, the rapidly growing megacities, general urbanization and industrialization across the globe, biofuels production, and the increasing effects of climate change. All have to be reckoned with as we struggle to ensure that there is enough water for people, food, and the environment. Sadly, the last of these, the environment, always loses out in such competition for water, and it is no coincidence that very significant declines in biodiversity, particularly in aquatic habitats, have occurred over the last few decades as development and consequent pollution have increased.

Given the pressures that water resources now find themselves under from all sectors of the economy and the environment, there is, in our view, an emerging world water crisis. While the severity of the water crisis may vary geographically, unless we find adequate solutions, its consequences—in the form of famine, environmental decline, social and public unrest, and migration—may impact both water-scarce and water-rich countries. Whether these ramifications result in water riots or wars, as postulated by some commentators, will depend on both the availability of innovative, science-based solutions to water scarcity and on governments' ability to reform water governance and ensure socially equitable outcomes for all water users. However, it is significant to note the comments of Professor Aaron Wolf of Oregon State University, who has pointed out that there has not been a significant war over water for approximately 4500 years and that water issues have often resulted in political agreement between peoples.[1]

As well as exploring the key factors that control water availability, scarcity, and use and to determine what increasing demand will do to global water resources, this book also looks at ways in which we can learn to use and manage our water resources better, with more innovation and with greater equity between all water users.

Some Water Facts

As we all go about our daily lives, we need water for drinking, cooking, washing ourselves and our clothes, removing our wastes, watering parks and gardens, and cleaning our houses, cars, and so on. To those of us living in "western" countries, water is available at the turn of a tap and at a cost, at least until recently, so negligible that we give it little thought. When we add up these uses, most of us manage on approximately 100–550 liters (26–145 U.S. gallons) per day depending on our climate and degree of profligacy (see Figure 1.1). However, billions of people worldwide still do not have access to safe drinking water and sanitation facilities. In June 2009,[2] three people were murdered following a fight over who should have first access to an illegal water connection in Bhopal, India. In other parts of the world, people often have to collect water from a standpipe in the street or even walk several kilometers to nearby rivers, lakes, or wells. In such cases, their survival is often based on usages of less than 50 liters (13 gallons) per day.

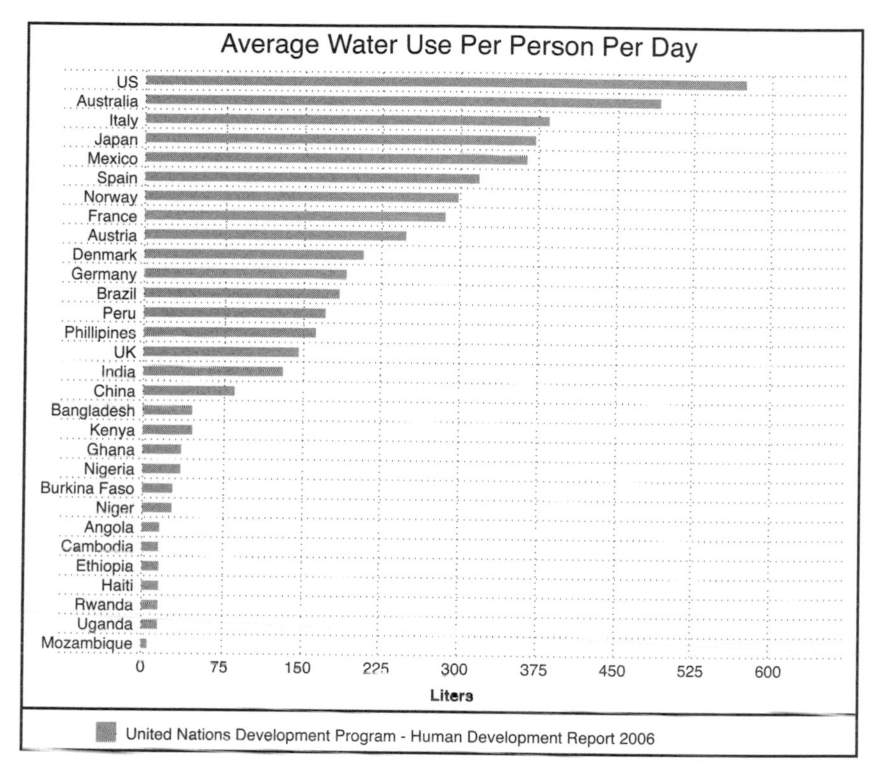

Figure 1.1 Per capita daily water usage

Source: UNDP, Human Development Report 2006 (http://hdr.undp.org/en/media/HDR06-complete.pdf) and www.data360.org

This book predominantly focuses on water use in agriculture. As a consequence some of the volumes involved become mind-boggling. Table 1.1 sets out commonly used metric and approximate comparisons with U.S. imperial units used for water measurement. The somewhat archaic acre-foot U.S. unit represents the amount of water that has to be applied to flood an acre to a depth of one foot. To put things in perspective, an Olympic size swimming pool of 50 x 25 x 2 meters (184 x 64 x 6' 7" feet) contains about 2.5 ML (660,000 gallons).

Lake Mead in the Western United States impounds a maximum volume of 35.2 km^3 (28,500,000 acre-feet). Throughout the book, while we try to keep numbers to a minimum, we quote the original unit of measurement first and then put its metric or imperial equivalent in parentheses.

Table 1.1 *Conversion of Metric to U.S. Imperial Measurements of Water*

Metric	US Imperial	
	Gallons	*Acre-feet*
1 Kiloliter (KL) = 1000 liters	**264**	**0.00081**
1 Megaliter (ML) = 1,000 KL or 1,000,000 liters	**264,000**	**0.81**
1 Gigaliter (GL) = 1000 ML	**264,000,000**	**810**
1 cubic meter ($1m^3$) = 1000 liters	**264**	**0.00081**
1 km^3 = 10^9 m^3 = 10^{12} liters or 1000 GL	**264,000,000,000**	**810,000**

Source: UNDP—Human Development Report, 2006

In 2002, many countries committed themselves to the UN's Millennium Development Goals that included ambitious targets of halving by 2015 the number of people without access to safe drinking water and sanitation. While the drinking water target may be met, there is increasing concern that the sanitation target won't be met.

The sheer task of meeting these targets is, furthermore, putting national government, donors, and aid agencies under considerable financial and other pressures, although the actual demand on water resources for drinking water is not particularly large. However, dealing with sewage, as more and more people get connected to sanitation systems in the world's growing cities, is going to be one of the world's greatest challenges in the forthcoming decades, particularly from the water quality, health, and environmental viewpoints. Many rivers in developing countries are now, in terms of pollution and poor water quality, similar to the putrid streams seen in cities like London in the nineteenth century.

In many ways, the goal that everybody should have access to safe drinking water and sanitation facilities has masked the fact that the real future demand on the world's freshwater resources will come from the use of water in agriculture for both food and fiber production (see Figure 1.2). Data from many countries with a significant agricultural base demonstrate that upwards of 70% of total water use goes to agriculture. This applies to developed areas such as California and Australia, just as much as to developing countries. When we examine the data for agricultural water use, drinking and sanitation water uses pale into insignificance. For example, a metric ton of rice requires up to 998,570,358 gallons (3780 ML) of water for its growth. A kilo or about 2 pounds of grain-fed beef requires about 2460 gallons (10,000 liters or 10,000 ML per ton). On average, every calorie consumed in our food requires a liter of water for its production, and this does not count water used by the food processing industries. Furthermore, "western" diets rich in animal products as compared with simpler (and healthier!) predominantly vegetable- and cereal-based diets require even greater volumes of water per calorie produced.

Figure 1.2 Agriculture is the largest water user often via irrigation. For example, irrigation has for centuries been the mainstay of Sri Lanka's agricultural economy. It is based on the construction of large reservoirs (tanks) which capture and store wet season rainfall.

Photo: Colin Chartres

Consequently, each person on the planet, if consuming a diet of 2500 calories per day, accounts for at least 660 gallons (2500 liters) of water requirement. Multiplied by 365 days per year this totals 241,056 gallons (912,500 liters, or nearly a megaliter). When we look at global population growth, which has climbed steeply in the last 50 years and is predicted to climb from about 6.7 billion in 2008 to about 9 billion in 2050 (see Figure 1.3), the future water requirement just to feed so many people, based on current levels of agricultural productivity, is approximately 3360 cubic miles (14,000 km^3).[3] The predicted extra 2 billion mouths equate to developing another 600–1440 cubic miles (2500–6000 km^3) of water resources. This equals at least another 25–50 enormous dams of the capacity (approximately 110 km^3) of the High Aswan Dam on the River Nile in Egypt. Herein lies the catch.

These vast amounts of water are not available or at least not available in the areas where we need them to produce food.

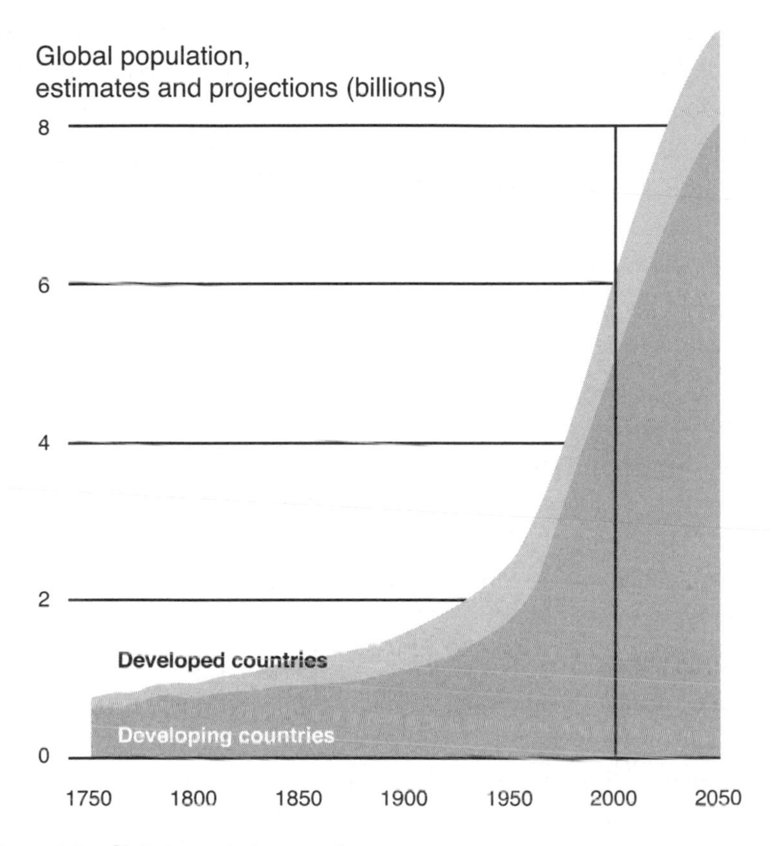

Figure 1.3 *Global population trends*

Source: Trends in population, developed and developing countries, 1750-2050 (estimates and projections). 2009. *Hugo Ahlenius, Nordpil, UNEP/GRID-Arendal Maps and Graphics Library*. Retrieved 07:19, April 22, 2010 from http://maps.grida.no/go/graphic/trends-in-population-developed-and-developing countries-1750-2050-estimates-and-projections.

Politically, one of the most pressing questions for many countries over the next 50 years will be whether they will be food-secure in the face of potential recurring world food crises. During the 2008 food crisis, although global food stocks reached worrying lows, there was still enough food available to feed everyone. The key problems were those of demand, driving prices above the reach of the poor and food stocks being geographically in the wrong place. Logically, for most countries, food security can be achieved by a combination of domestic production and imports, but as was observed in the 2008 food crisis, logic often flies out the window in the face of adversity. Some food exporting countries adopted policies that were clearly driven by fear and stopped exports. Other countries froze grain prices, which was detrimental to poor farmers, encouraged black marketeering, and did little to aid the urban poor. So some critical needs for many countries are policy settings that encourage domestic production via increasing productivity, recognizing the benefits of importing "virtual water" via trade in food, and enabling poor farmers to benefit from rising prices so that they in turn can invest in more productive systems. Irrespective of individual countries' different policy responses to food crises, it seems inevitable that increasing pressures for food security will play a big role in determining water policy and management responses.

Much of the world is already water-scarce. The availability of water and access to water will be major issues for economic development and for the livelihoods of the poor, given that they often suffer most when resources are scarce. Water scarcity can be described as being physical or economic in nature (see Figure 1.4).

Physical water scarcity results from the allocation of virtually all available water supplies, leaving nothing for additional use or for the future or for the environment.

Water scarcity has become a reality for many regions. Much of south and west Asia, China, the Middle East, Northern and Southern Africa, Southern Australia, and the Southwestern United States are in this category. Physical water scarcity will put increasing pressure on water planners and managers to develop ways to better manage their existing water resources, to increase the productivity of water, and to develop "new" sources of water, that is, "reuse" of wastewater. Many countries have already seen water users turn to groundwater often not cognizant of the high degree of connectivity between groundwater and surface water.

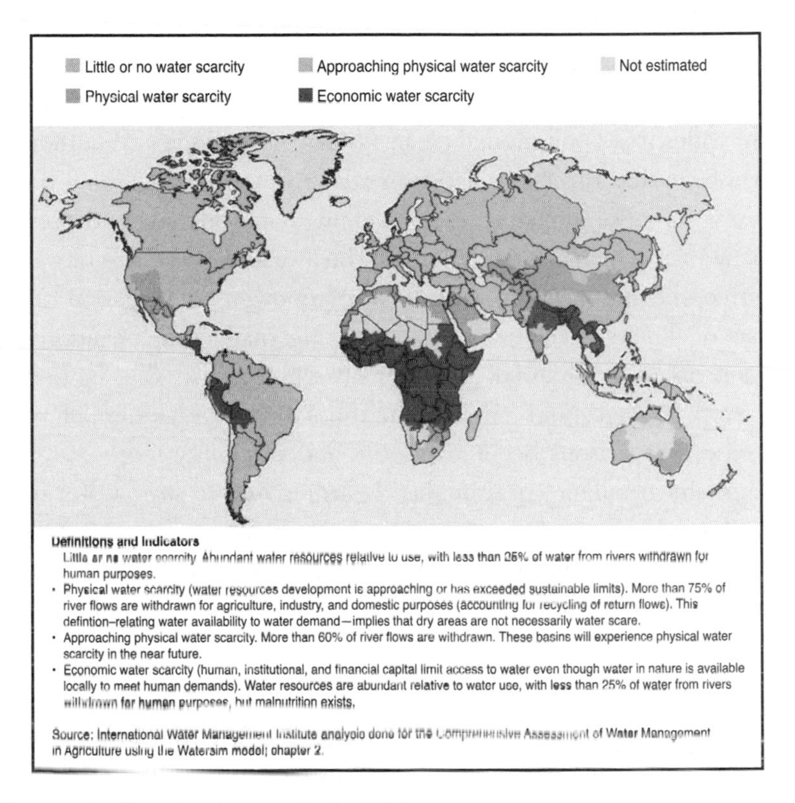

Figure 1.4 Global water scarcity in 2000

Source: Comprehensive Assessment of Water Management in Agriculture

There are, however, many areas in the developing world, in particular in sub-Saharan Africa and parts of Southeast Asia, where there are still available water resources, but development and use of these resources has been constrained by lack of capital investment or appropriate institutions to support the use of that capital. The resulting "economic" water scarcity has major ramifications for the poor and economic development in general, and its solution has the potential to bring global benefits and reduce stresses on other water-scarce areas. The issue of insufficient infrastructure development also relates to limited investment in wastewater treatment facilities and the consequent widespread pollution of clean surface water bodies. Whether in areas of physical or economic water scarcity, a critical factor for the future will be the impact of climate change and ongoing and potentially increasing climatic variability on the availability and use of water resources be it for drinking water, hydropower, or irrigation. The impact of climate change will vary depending on geography and scale. In some areas, total rainfall and intensity will increase, causing flooding, crop damage, and erosion. In other areas, total rainfall may decrease, wet seasons become shorter, and variability more extreme with greater frequency of droughts. Learning how to store water better and providing supplementary irrigation to make up for erratic rainfall supplies will be the key to overcoming these challenges. Climate change and potential adaptation are discussed in more detail in later chapters.

So, the preliminary facts and figures presented here demonstrate very considerable cause for concern as to where we are going to get the water that will be needed to sustain us in the future. It is not appropriate to brush off this concern by saying that I live in a country well-endowed with water resources, and this is therefore not my problem. Local and regional problems these days tend to escalate politically and physically across the globe. Water scarcity and food scarcity in one place, just like political oppression, manifest themselves in the form of political and social unrest and illegal migration, which can become of major concern to neighboring and receiving countries, respectively. The following chapters explain in more detail the nature and dimensions of water availability and use at the global level and look at their severity and likely future impact both regionally and internationally. From our perspective, although we believe that a water crisis is imminent, we do not believe that panic and kneejerk solutions are the way forward. The book is written around the key concept of using scientific and socio-economic evidence as the basis for reform and management action in the water sector. It also focuses on the need to look at water issues in an integrated fashion as opposed to dealing with water supply, sanitation, agricultural, and environmental water separately. What is for sure is that given the pressures facing water supply and demand and their management, we can't continue using practices developed, in some cases, over 150 years ago. We have to change the way we do "business" in the water sector once and for all. As well as looking at the problems, we explore the right and wrong ways that they can be tackled and put considerable focus on mixtures of low tech, high tech, socio-economic and governance solutions to the world's water problems.

chapter two

From Abundance to Scarcity in 25 Years

"All the water that will ever be is, right now."

National Geographic, October 1993

Over the last quarter century there has been a fascinating transformation in terms of the way water is viewed by the global community. In most of the Western world, prior to the last couple of decades, water was of high quality; readily available to agriculture, domestic, and industrial users; and provided at very low cost. Over time, and specifically because of the advancements in technology and demands of human activity, many countries have now shifted to a situation where available water resources are close to their sustainable limit of extraction, and new users wanting access to water are creating competition in the "water market place." This means that we are witnessing a change from a situation where rivers that used to flow with large volumes of water into the ocean do so with much lower flow or only intermittently. Technically, river basins where this occurs are said to be transforming from "open" to "closed" basins. At this stage we move to a situation where further development of water resources shifts from the physical debate about how to make water available to a political debate about who should get *how much* water. As a consequence, the range of skills and inputs needed by governments and the private sector to manage water resources is also shifting, from being predominantly focused on engineering to a mix of science, engineering, economics, and social sciences. The topic of water is moving into the public arena as a major issue confronting society in the same way as are the topics of energy, climate change, and food security. In fact, as will be shown later, all of these issues are inexorably linked.

Existing thinking on water management is being forced to change. The eloquent description that follows reveals how water development and management was and still is approached:

> "Knowledge about water has been overwhelmingly dominated by natural and technical sciences bent on harnessing, controlling or regulating water regimes for economic development. Engineering skills have historically been concentrated in powerful colonial or national state water agencies. The 20th century has witnessed the apogee of the so-called "hydraulic mission," a period in which engineering approaches at dominating nature fully blossomed, leaving behind a balance of 50,000 large dams and 280 million hectares (692 million acres) of irrigated land, on which a substantial part of humankind's food and energy production is predicated."[1]

In many countries in the developed world, particularly the emerging giants of India and China, the "hydraulic mission" was the focus of new modern nations and can be credited with supporting the green revolution and lifting millions out of poverty and hunger. However, in many African countries and in several other very poor countries, things are not the same. A prolonged lack of investment in adequate water infrastructure development has affected the abilities of these countries to grow enough food and also to support other economic activities, thus creating a condition of economic water scarcity and worsening poverty.

Other major concerns raised in the focus on the "hydraulic mission" of the past decades were the environmental consequences of water development including serious declines in biodiversity and environmental degradation (see Figure 2.1). In extreme situations, over-exploitation has led to the destruction of major water systems, exemplified by the drying of the Aral Sea. It is now clear, based on cost estimates for righting these mistakes, that attempts to rectify the enormous impact and cost of these environmental problems will be far, far greater than would have been the cost of establishing built-in concepts of sustainability into the original schemes. Water experts and planners are now being pressured to think differently about water development, developing a deeper understanding of the balance between ingoing capital investment in infrastructure and its replacement and the costs to the environment and the services it provides.

Figure 2.1 Major landscape change as exemplified in the change from natural forest to plantation agriculture (in this case, tea in Sri Lanka) have had profound changes on landscape hydrology, erosion, sedimentation, and biodiversity.

Photo: Colin Chartres

The presence of water resources management on the mainstream political agenda of many countries has provided lessons for how water resources have been used and exploited often to the detriment of the source. In this chapter we look to some of these examples, starting in the Middle East, where obvious physical scarcity was worsened by management decisions that did not consider sustainable use. This journey will start with the Jordan River and move through India, Australia, and the United States, exploring the histories of some of the world's most critically water-stressed areas.

The River Jordan

You may have heard of the agricultural "miracles" that have occurred in the deserts of Israel and to a lesser extent its neighboring countries. Israel leads the world in some aspects of high-tech irrigation technology. It also has exemplified, to the greatest extent, the maxim of an ancient Sri Lankan king, Parakramabahu, who famously declared, "Let no drop of water flow to the sea unused by man."[2] However, the environmental consequences of such practices may be unsustainable and extremely costly. In the case of the River Jordan, the cost of over extraction and overuse of water by Israel and Jordan, is leading to the potential death of the Dead Sea and the promotion of a several hundred-billion-dollar Red-Dead Sea canal to save it. Just how did this happen?

The Lower Jordan River Basin is fundamental to the survival of the country of Jordan, given that it receives 80% of the national water resources.[3] Eighty-three percent of the total population lives in the basin, as well as it having most of the country's industry and 80% of the irrigated agriculture. International conflict has also affected the basin, given that its headwaters now lie partially in the Golan Heights, which is disputed by Syria and Israel. These headwaters drain into Lake Tiberius, which is now used as a freshwater source by Israel. The outflow of water from this lake is virtually blocked and contributes little to the flow in the Lower Jordan.

Consequently, the Lower Jordan River depends upon the Yarmouk and Zarqa Rivers that originate in the northeast of Syria and from several ephemeral streams known as *wadis*. Prior to water resources development in the Jordan Valley, the flow of the Jordan into the Dead Sea was estimated, by several hydrologists[4] at between 1,100 and 1,400 million cubic meters per year (Mm^3/yr) (892,000–1,135,000 acre-feet per year). As explained next, by the mid-1970s this figure had reduced by half, and by the early 2000s, this was approximately halved again. Originally, the surrounding highlands to the east and west were covered by forests composed mainly of Mediterranean conifers, but these have subsequently been deforested and are now rangelands with occasional olive and stone-fruit trees. Further east is an upland plateau that is used for cereal cultivation in the wetter zone near the mountains and for nomadic pastoralism in the drier east. Several large urban centers including Amman, the Jordanian capital are found on this plateau.

Most of the rain that falls in the in the Lower Jordan River Basin is evaporated or *evapotranspirated* (water that is transpired by plants or evaporated from surfaces) by irrigation and rainfed crops. Only a small percentage (5%) gets into the river, although a modest amount (15%) percolates into groundwater. The impact of the agricultural and urban development of Jordan and of Israel taking more water from Lake Tiberius on the Lower Jordan Valley are shown graphically in Figures 2.2, 2.3, and 2.4.

Population growth was the major driver of the changes observed. Between 1948 and 1975 the population of Jordan increased dramatically from about 450,000 to 2 million people. The rate of increase was boosted by two waves of refugees, following the creation of the state of Israel and secondly after the Six-Day War in 1967. This increase in population led to the international community financing large facilities for the development of irrigated agriculture as part of a wider socioeconomic development process. In 1977 the Israelis also raised the level of Lake Tiberius, reducing its outflow to 70 Mm^3/yr (56,750 acre-feet per year). Urban expansion also started to draw heavily on groundwater, and agricultural development on the Yarmouk tributary further reduced inflows into the Lower Jordan River.

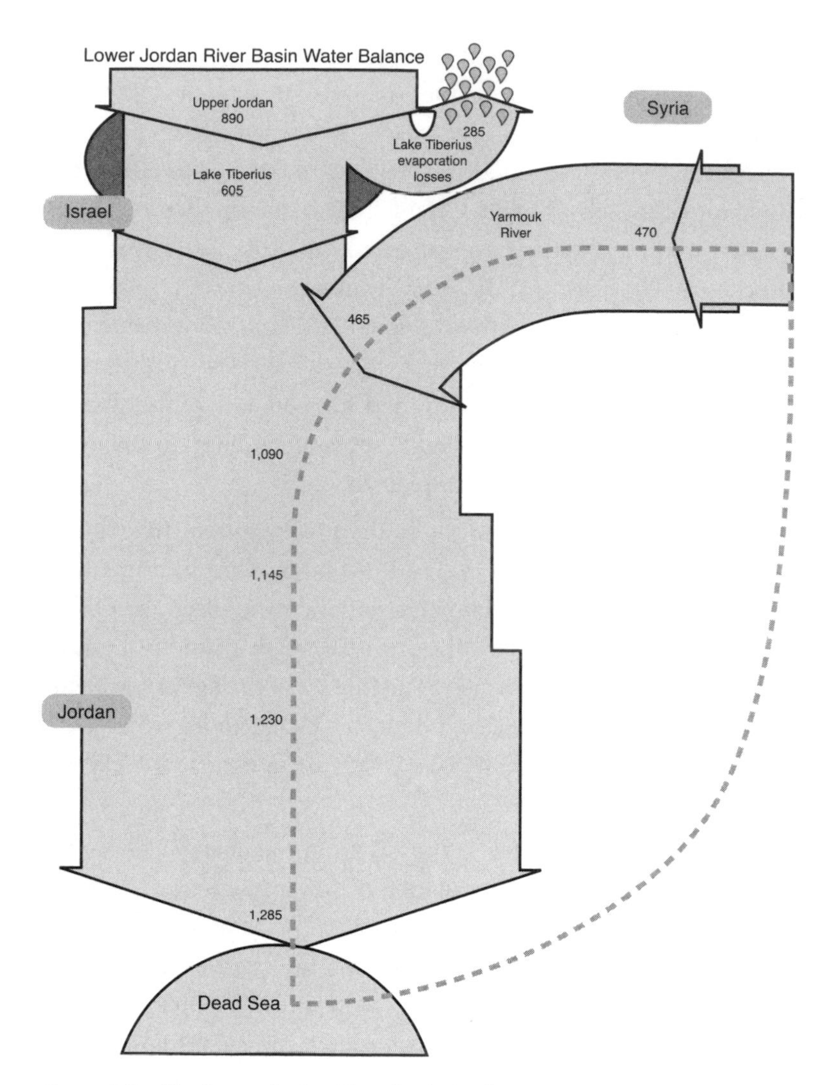

Figure 2.2 The Lower Jordan River Basin in the 1950s

Source: Adapted from Courcier, et al. (2005)

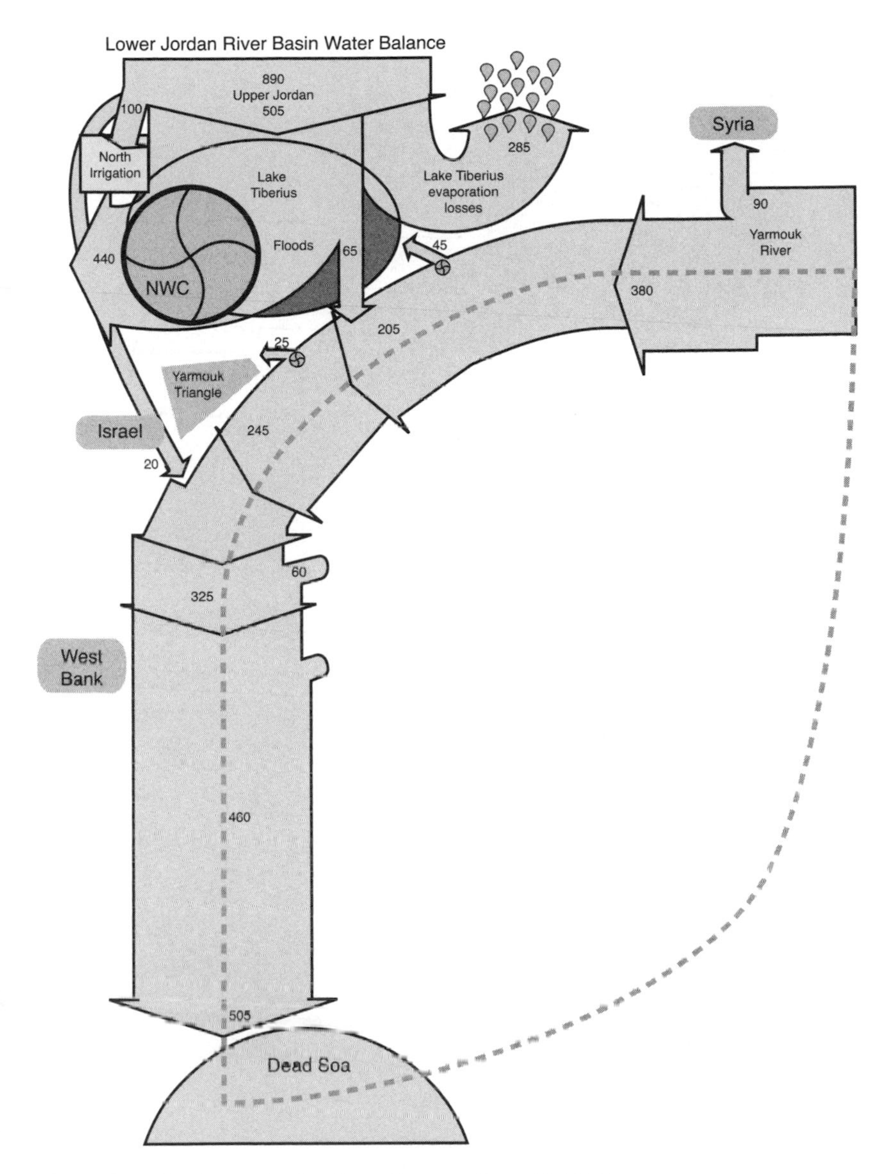

Figure 2.3 The Lower Jordan River Basin in the mid-1970s

Source: Adapted from Courcier, et al. (2005)

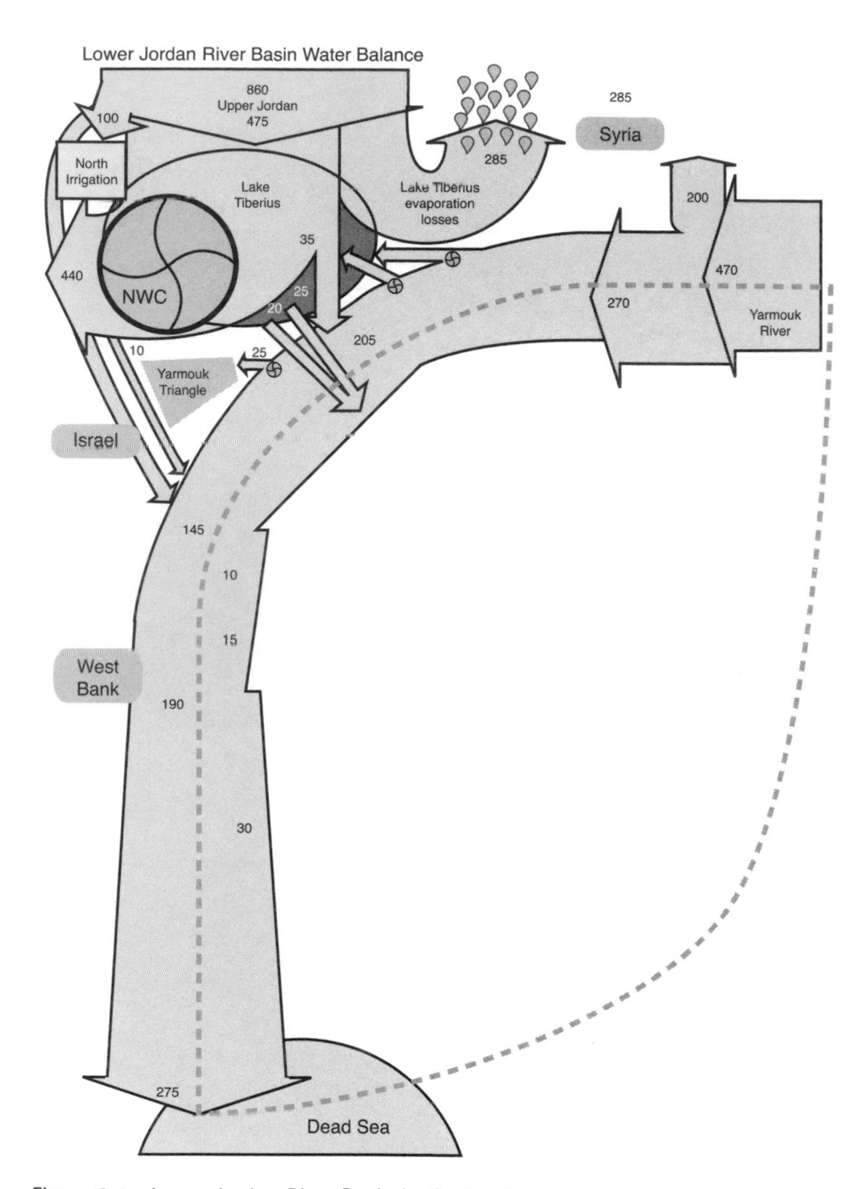

Figure 2.4 Lower Jordan River Basin in the 2000s

Source: Adapted from Courcier, et al. (2005)

Agriculture and other development continued apace through the 1980s and into the current century, much of it financed by international aid. Improvements in irrigation technology allowed many small entrepreneurial farmers to produce crops year-round with good profitability. Unlimited access to groundwater resources on the desert plateaus also fueled a similar boom in agricultural development with the emergence of many large olive plantations. The impacts of all this greater water extraction and use reduced the flow into the Dead Sea further to 275 Mm^3/yr (223,000 acre-feet per year) with an accompanying decline in water quality.

By the mid-2020s a further diminution of flow and quality is expected, with Dead Sea inflows reducing to 170 Mm^3/yr (138,000 acre-feet per year). Under this scenario, it is probable that diversions to agriculture will remain stable, but salinization of water and soils will become an increasing issue. Furthermore, demand for water in the region is such that the growing population centers will be supported by increasing extraction of groundwater, some of it fossil, and the desalinization of saline discharges into the Jordan and other brackish and saline water resources.

The ultimate result of development, population growth, and Israeli water diversions in the Lower Jordan Valley is that flows out of Lake Tiberius and ultimate discharges into the Dead Sea have reduced by approximately eight-fold from pre-1950 (see Table 2.1). This demonstrates quite vividly the kind of pressure that water resources are facing around the globe, particularly in relatively dry environments.

The possible consequence of utilizing virtually all the flow in the Lower Jordan and the increasing demand for water in the region is the proposition of constructing the Red-Dead sea canal, which would flow for a distance of 180 km (112 miles), using the drop of 400 m

(440 yards) between the Red Sea and the below sea level Dead Sea. Thus the Dead Sea level could be stabilized by bringing in a volume of water similar or slightly higher than that originally flowing into the Dead Sea from the River Jordan. A further significant volume of saline water would also be used to supply the main cities of Jordan. The costs of building the canal and of the desalinization of the Red Sea water are enormous and may exceed $5 billion, but the project is receiving serious consideration from Jordan and its neighbors as well as international funding agencies. To what extent hydropower generated from the gravity flow of water in the canal can be used to power the desalinization process is as yet uncertain. In conclusion, the question has to be asked as to whether the overexploitation of the water resources in the Jordan Valley by both Israel and Jordan has been beneficial. Clearly, the agricultural development of the region has stimulated economic development in both countries, but at great cost both to the environment and now in terms of the need for major and extremely costly infrastructural development. As will be discussed later in the book, an alternative pathway for water-scarce countries is to conserve water for the highest value uses (which may include some fruit and vegetable production) and import staple food commodities (virtual water) from water-rich countries.

Table 2.1 *Flows Out of Lake Tiberius and Water Reaching the Dead Sea in the Jordan Valley Pre-1950 to 2020s*

	Pre-1950s Mm^3	Mid-1970s Mm^3	Early-2000s Mm^3	Predicted Mid-2020s Mm^3
Flows out of Lake Tiberius	605	70	70	70
Discharges into the Dead Sea	1,370	505	275	170

The South Asian Groundwater Phenomenon

Over the last half century or less, India has seen a phenomenal change in the way in which water is used in agriculture. South Asia (India, Pakistan, and Bangladesh) emerged from the colonial era with the world's largest centrally managed canal irrigation infrastructure.[5] However, if we look carefully at the region today we see that much of this surface irrigation system has been replaced by the use of groundwater pumped from literally millions of wells, and from boreholes drilled in a largely unregulated fashion. While this change has allowed many small farmers access to water when they require, it is clearly not without risk because of the pressure it places on the region's groundwater resources.

These pressures are exemplified by a recent study[6] that confirms, using the NASA Gravity Recovery and Climate Experiment satellites, that groundwater is being depleted at a mean rate of 4.0±1.0 cm yr (1.6±0.4 inches per year). Under the Indian states of Rajasthan, Punjab, and Haryana (including Delhi). During the study period of August 2002 to October 2008, groundwater depletion was equivalent to a net loss of 109 km^3 (88 million acre-feet) of water, which is double the capacity of India's largest surface-water reservoir. Annual rainfall was close to normal throughout the period, which shows that the other terrestrial water storage components (soil moisture, surface waters, snow, glaciers, and biomass) did not contribute significantly to the observed decline in total water levels. The study concluded that although the observational record is brief, the available evidence suggests that unsustainable consumption of groundwater for irrigation and other human and domestic uses is likely to be the cause. If measures are not taken soon to ensure sustainable groundwater usage, the consequences for the 114 million residents of the region may include a reduction of agricultural output and shortages of potable water, leading to extensive stresses that could impoverish the population.

Tushaar Shah, one of India's leading groundwater experts, describes the change from surface water, gravity-based irrigation to pump irrigation using groundwater as "irrigation anarchy." He explains that the previous highly regulated surface water irrigation system has been overtaken by individuals behaving in an "atomistic" (hundreds of thousands of farmers behaving independently in an uncoordinated manner) fashion. Just how striking the change has been is demonstrated in Figure 2.5, which shows the incredible growth in pump irrigated areas between 1800 and 2000. During this period, the proportion of irrigated area to the total area sown climbed from 10% to over 53%.

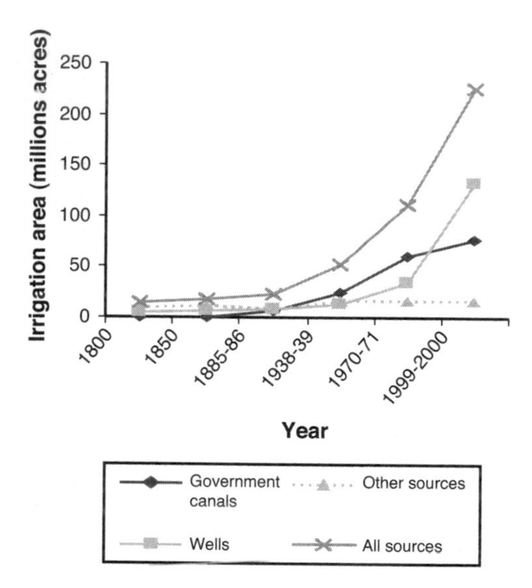

Figure 2.5 Transformation of South Asian irrigation in India, Pakistan, and Bangladesh 1800–2000: net area (million acres) by irrigation source

Source: Shah (2009)

The transformation of irrigation in South Asia is marked by three indicators:

1. **Vastly increasing numbers of water extraction mechanisms.** Shallow wells and tubewells (which are a type of water well in which a long 5 to 8 inch-wide stainless steel tube or pipe is bored into the underground reservoir) and their pumps are used to lift water from ponds and rivers. By 2007, it was estimated that there may be 27 million such structures in India alone. This figure has increased from around 200,000 in 1960.
2. **A rapid increase in the irrigated area dependent on groundwater.** In 1950–51 water extracting mechanisms accounted for the irrigation of 11.9 million ha (29.4 million acres). This figure had risen to 53 million ha (131 million acres) in 2006.
3. **The changing ratio of surface water irrigated area to groundwater irrigated area**, as shown in Figure 2.5.

Shah argues that the growth in groundwater irrigation in South Asia is not the result of pump technology simply allowing otherwise suitable land in dry regions to be irrigated, as has been the case elsewhere in the world, but that in South Asia the growth has been spurred by scarcity of farmland more than scarcity of water. Irrigation geography in South Asia has been transformed, taking the water away from the controllable areas of tanks and canals and spreading it throughout the continent, even into areas where geological conditions (fractured rock aquifers) are generally considered unreliable for sustainable water supply. He also lays out a number of supply-side theories as to why pump irrigation is preferable to gravity-fed systems (it allows areas out of command [that is, higher in the landscape] from tanks and rivers to be irrigated, enabling farmers to get water exactly when they want it, and has worked especially well when it is supported by cheap electricity), which has also prompted a lot of groundwater use. He also argues that institutional and regulatory systems for surface water systems are weak in South Asia, and farmer participatory irrigation management approaches in the surface systems have not worked well. Given the combination of land scarcity, supply-side failures in the canal system, and the advent of appropriate technology, pump irrigation has exploded across the region. The only other area that has seen similar growth in the use of predominantly groundwater for irrigation is the North China Plain, where groundwater use commenced in the 1970s and took off in the 1990s.

Shah contends that the strong links that have developed between pump irrigation and agrarian poverty in South Asia are the "hallmark of the water scavenging economy"[7] that has come to dominate the region. In its early stages, pump irrigation created value by enhancing land use intensity. Now it also actually creates new opportunities and livelihoods by allowing more diversification of agriculture, including the production of higher value crops.

While the changes in irrigation in South Asia and China are physically remarkable, both in terms of the amount of area impacted and socio-economically in terms of the way farmers have responded to changing drivers, they are also a major concern from a hydrogeological and food production point of view. This is because there is limited information available to assess the long-term sustainability of pump irrigation systems that draw primarily on groundwater. Groundwater accumulates below the earth's surface when there is a surplus of water that infiltrates the soil and is not then used by plants or evaporates. This water moves very slowly under the influence of gravity and under natural conditions may discharge in topographic low spots into lakes and rivers. While there is some fossil groundwater that accumulated millions of years ago and then got trapped because of changing geological conditions, the majority of groundwater exploited for human use are intimately connected to surface water. So when they are drawn off for human uses, such as irrigation and domestic water supply, a critical issue is whether they are being utilized at the rate of annual recharge or more. If the latter, then the groundwater table will drop, and other subtle changes, such as the reduction of flow into rivers and lakes, may occur, which may have an impact on dry season river flow and ecosystems. So ideally, each known groundwater resource should have what can be defined as a sustainable yield, which may not be the same as annual recharge, but which will approximate some average rate of recharge or maintain the aquifer at a level above defined thresholds.

Experts have warned that a quarter of India's food harvest is at risk if the country fails to manage its groundwater resources properly.[8] Aggregate data for India suggest that annual groundwater use is only 50% of the total annual utilizable groundwater. However, this hides regional differences. Shah defines four stages of groundwater development. The first, the "green revolution," occurred in the 1960s and '70s when new crop varieties combined with fertilizer access and, in some cases, irrigation saw major increases in food production, productivity for staple crops like rice and wheat; the second, a groundwater-based agrarian boom, then early symptoms of groundwater overdraft, and the final stage is the decline of groundwater socioecology, resulting in an "immiserizing" impact. Several regions of India have reached this fourth stage, which leads to migration, decreasing food production, and suffering of those who have not made a transition out of the agricultural economy. In South Asia these problems are more likely outside the areas of command (areas around the dam that benefit from irrigation water of the surface irrigation systems but are nonetheless of major concern. The most critical issue, however, is that with weak institutional and regulatory structures, there are few ways that governments can step in to try to manage water use and thus decrease overall water withdrawals. Unless governments in South Asia make the difficult decision to address unsustainable use of groundwater, many areas and many people will inevitably suffer major consequences of groundwater depletion in the next 20 to 30 years. Given pressure on water resources in the region, governance reform has to be a critical feature of the political economy of the twenty-first century. Such reforms are discussed in later chapters.

The Murray-Darling Basin

While Australia has many agricultural success stories, for the couple of hundred years since European settlement, it has been bedeviled by agricultural developments that have failed because of a misunderstanding of its climate, soils, and water resources. The story of the Goyder line in the state of South Australia illustrates how during the 1870s wheat growers pushed further north to establish farms against the advice of the Surveyor General, George Goyder, who warned that the droughts of the previous decade were likely to return. They persisted in their plan, and "forced to retreat, the settlers left behind a degraded landscape littered with ruined buildings that can be seen to this day."[9] The farmers of that period had little understanding of the variability of Australia's climate, which still confounds their descendants today, where "a similar struggle between biophysical realities and human ambition is underway in the Murray-Darling Basin where the process of landscape and stream modification has proceeded apace in recent decades largely oblivious of the need for caution or the possibility of threshold changes to its ecological systems."[10]

Located in the southeast region of Australia, the Murray-Darling Basin covers 1,061,469 square kilometers (409,835 square miles), equivalent to 14% of the country's total area (see Figure 2.6). East-west, the basin extends 1250 km (775 miles) and is 1365 km (850 miles) from north to south, in latitudinal terms, from 24°S to almost 38°S.

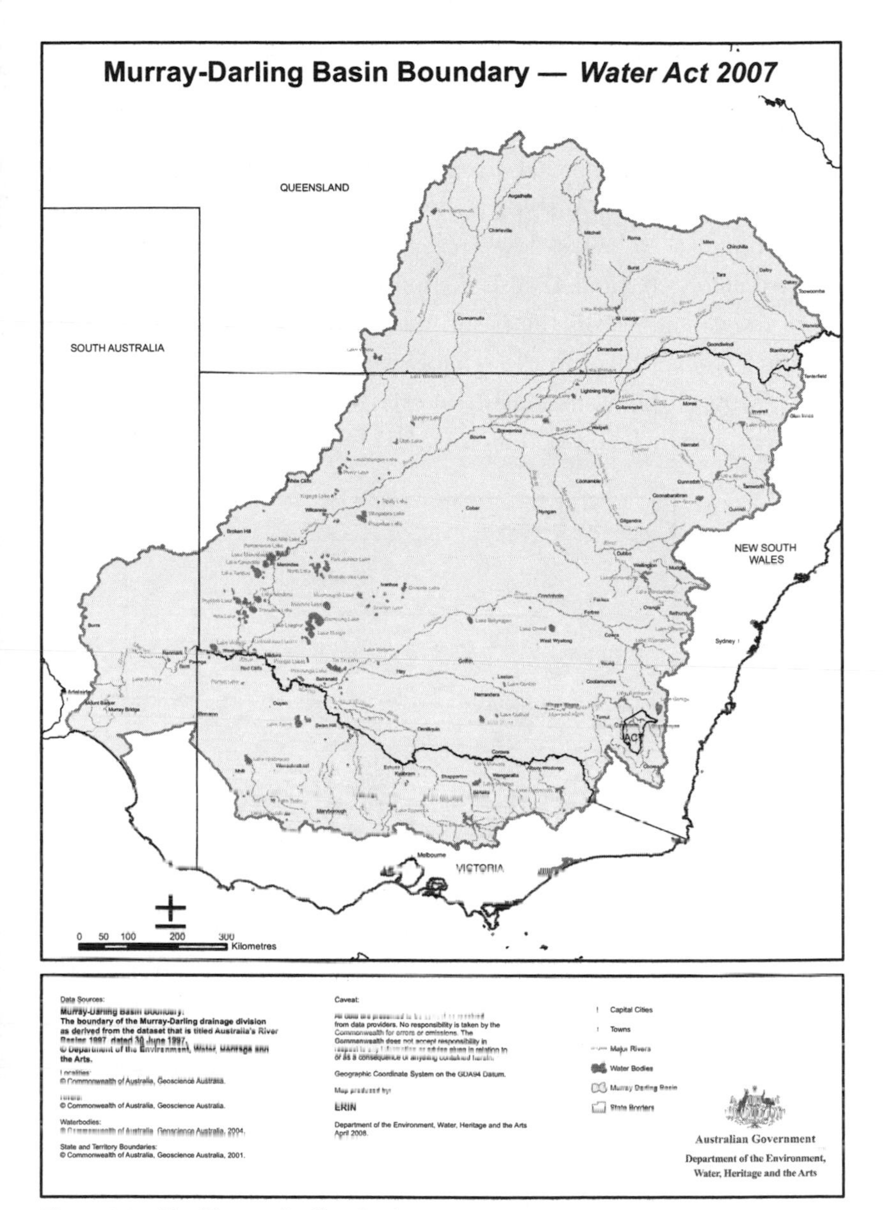

Figure 2.6 The Murray-Darling Basin

Source: Department of the Environment, Water, Heritage, and the Arts, Government of Australia, April 2008

To the east and south, the Great Dividing Range forms the limit of the basin, including Australia's highest country, where the majority of the basin's rainfall occurs. However, rainfall across the entire basin has always been somewhat irregular, and droughts and floods are characteristic of the Australian environment (see Figure 2.7). The ratio of highest to lowest flows in the Murray and Darling Rivers are approximately 16:1 and 4700:1, demonstrating the increasing variability of the climate across the basin. By way of contrast, the same ratio for the Potomac River in the reliably watered Eastern United States is 4:1. Most of the basin consists of extensive plains and low undulating areas, mostly below 200 meters (655 feet) above sea level.

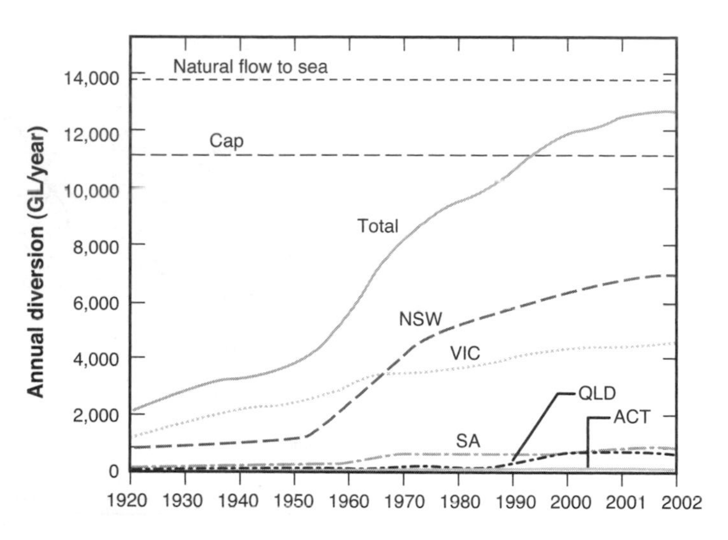

Figure 2.7 Growth of total water use in the Murray-Darling Basin 1920–2000 (1 GL is a billion liters or approximately 810 acre-feet)

Of greatest extent are the vast plains, the Darling Plain in the north, drained by the Darling and its tributaries, and the Riverine Plain in the south, drained by the Murray and Murrumbidgee and their tributaries. In terms of agricultural production, the basin accounts for about 40% of Australia's agricultural output, much of it from irrigation. On average, 14,000 gigaliters (GL) (over 11 million acre-feet) of the Murray-Darling flow used to drain into the sea on an annual basis. The history of the development of water resources in the Basin would be similar to that in many other basins around the world. Initially in the nineteenth century, the basin's rivers were navigable and used for transportation However, as the irrigation industry was developed during the twentieth century, particularly in the period after the Second World War, increasing water extraction, coupled with periods of low flow, considerably reduced the extent of navigability. Development included the building of several large dams on the major upstream tributaries of the Murray, Murrumbidgee, and Darling and a significant storage (the Hume Dam) on the Murray itself. This significantly increased storage in the basin to levels that are in the region of several thousand cubic meters per inhabitant. As has been demonstrated in recent years, even this has not been enough to deal with the record-breaking drought of the early part of the twenty-first century. By the 1980s, virtually all the basin's water had been allocated to users, the majority of whom were irrigators. By this time significant environmental stress was becoming apparent across the basin. This was manifest in the form of devastating land degradation, including dryland and irrigation salinization, soil acidification, erosion, and a major decline in aquatic and terrestrial biodiversity. In particular, water quality in the Murray was deteriorating in the face of higher salt concentrations, which was of major concern because the river provides much of the city of Adelaide's drinking water. Given these significant problems, the Murray-Darling Basin Commission developed a salinity and drainage strategy and following spectacular but toxic algal blooms in the Darling River in the summer of 1991–92, agreed to place a temporary cap on further extraction over 1993–94 levels in 1995. This was made permanent in 1997 and complemented by a salinity credits trading scheme aimed at reducing the increasing salinity levels in the river. In spite of these actions, by the 1990s total flows to the ocean were about 20% of predevelopment flows, and the Murray actually stopped flowing into the ocean on a number of occasions, necessitating maintenance of an artificial channel by a dredger.

Since the turn of the present century, the southeast region of Australia has been under severe drought conditions. Water allocations have been a small percentage of full entitlements, rice production has declined significantly, and the irrigation communities have been put under severe economic stress. Significant environmental assets are also on the verge of collapse in the river valleys and associated wetlands. These occurrences have triggered some innovative policy responses from the government, including the separation of water and land rights, which freed up the development of a market for water entitlement and allocation trades. While this market does have some constraints imposed by the Australia states involved, it has led to significant volumes of water being traded from low to high-value crops and will also facilitate the purchase of water for environmental uses. To achieve these outcomes the government established the National Water Initiative in 2004, which was administered by the Australian Government Water Fund, which in turn was responsible for funding infrastructural investment and knowledge-based programs across Australia. It also played a very significant role in policy reform. However, in the Murray-Darling Basin, it was also becoming clear that the old River Murray Commission and its successor the Murray-Darling Basin Commission that had operated for approximately 80 years, had a constitution and membership (that is, the states and Commonwealth of Australia) more suited to development and allocation of the basin waters than management of a reducing volume of water. Essentially vested interests at the state level became the sticking point against necessary reform. Consequently, in 2007, the Australian Government, with the agreement of the states, took control of the Murray-Darling Basin and established a new Murray-Darling

Basin Authority, which reports to the relevant Australian Government Minister. This takeover was only achieved by the Australian Government putting several billion dollars on the table to improve water use efficiency at a system and farm level and to invest in improving environment flows and protecting high-value environmental assets. Even with current and future investment of approximately 12 billion U.S. dollars of investment, the fundamental decrease in water availability, particularly in the southern basin, looks like it will result in a significant number of irrigators in the basin going out of business in the next few years. Recent studies by the CSIRO Australia, suggest that the rainfall declines in the southern basin may well be the consequence of anthropogenic climate change, although differentiating this from climate variability is complicated. More detail is provided on the Australian experience of water reform later in the book.

The Colorado River

"The Future Is Drying Up," an article by Jon Gertner from the October 21, 2007, issue of *The New York Times Magazine* stated:

> "In the Southwest this past summer, the outlook was equally sobering. A catastrophic reduction in the flow of the Colorado River—which mostly consists of snowmelt from the Rocky Mountains—has always served as a kind of thought experiment for water engineers, a risk situation from the outer edge of their practical imaginations. Some 30 million people depend on that water. A greatly reduced river would wreak chaos in seven states: Colorado, Utah, Wyoming, New Mexico, Arizona, Nevada and California. An almost unfathomable legal morass might well result, with farmers suing the federal government; cities suing cities; states suing states; Indian nations suing state officials; and foreign nations (by

treaty, Mexico has a small claim on the river) bringing international law to bear on the United States government. In addition, a lesser Colorado River would almost certainly lead to a considerable amount of economic havoc, as the future water supplies for the West's industries, agriculture and growing municipalities are threatened. As one prominent Western water official described the possible future to me, if some of the Southwest's largest reservoirs empty out, the region would experience an apocalypse, 'an Armageddon.'"[11]

Like the Murray-Darling Basin, the Colorado Basin in Southwest United States and Mexico, is also fast becoming a "closed" basin in that its water only intermittently flows into the Gulf of California due to the vast diversions for agriculture in California. Also, like the Murray-Darling Basin, the Colorado supplies much of the irrigation and drinking water for the Southwest. The Colorado River supplies water to seven states (Arizona, Colorado, California, New Mexico, Utah, Wyoming, and Nevada) in addition to Mexico. Since the early 1900s, water distribution rights have been in contention. And starting in 1922, the distribution and management of Colorado River water has been governed by a complex body of laws, court decrees, compacts, and an international treaty.

The Colorado originates in the Rocky Mountains and flows 1450 miles (2330 km) to the sea through a descent of 9000 feet (2740 m) and includes a number of major storages and hydropower plants. Almost 90% of the river's flow is diverted, with the majority (90%) used by agriculture. A key feature of the Lower Colorado is the All American Canal, the largest irrigation canal in the world that diverts water into the Imperial Valley. In 1905, an engineering disaster led to the river flowing uncontrolled through a diversion canal and the formation of the Salton Sea. This inland sea has progressively become more saline, and a number of schemes have been proposed and some of them implemented to reduce salinity levels and preserve the sea as a significant bird habitat.

Original agreements with Mexico dealt with the quantity of water in the Colorado that the United States agreed to deliver. The Colorado River Compact of 1922 agreed to the allocation of the Colorado's waters as shown in Table 2.2 with a further 1.5 million acre-feet (1850 Gigaliters) allocated to Mexico in 1944.

Table 2.2 *Water Allocations in the Colorado Basin as Determined by the 1922 Compact Among the Seven Basin States**

State	Water Allocation (Million Acre-feet/Year)	Percentage
Upper Basin		
Colorado	3.88 (4,790 GL)	51.75
Utah	1.73 (2,135 GL)	23.00
Wyoming	1.05 (1,295 GL)	14.00
New Mexico	0.84 (1,035 GL)	11.25
Arizona	0.05 (60 GL)	0.70
Lower Basin		
California	4.40 (5,430 GL)	58.70
Arizona	2.80 (3,455 GL)	37.30
Nevada	0.30 (370 GL)	4.00

*Approximate gigaliter (GL) equivalents are shown in parentheses.

In 1928, the U.S. Congress required California to "irrevocably and unconditionally" agree to limit its annual use of Colorado River water to 4.4.million acre-feet (5430 GL). Congress made the state's agreement a condition for building the Hoover Dam. However, California's use of the Colorado increased above this stipulated figure over the years, much to the chagrin of the other basin states. In recent years, the compact has become the focus of even sharper criticism, in the wake of a protracted decrease of rainfall in the region. Specifically, the amount of water allocated was based on an expectation that the river's average flow was 16.4 million acre-feet per year (642 m^3/s).

Subsequent tree ring studies, however, demonstrated that the long-term average water flow of the Colorado is significantly less.[12] Estimates have included 13.2 million acre-feet per year (516 m^3/s), 13.5 million acre-feet per year (528 m^3/s), and 14.3 million acre-feet per year (559 m^3/s). Many analysts have concluded that the compact was negotiated in a period of abnormally high rainfall and that the recent drought in the region is in fact a return to historically typical patterns. The decrease in rainfall has seen widespread dropping of reservoir levels in the region, in particular at Lake Powell, created by the Glen Canyon Dam in 1963.

During the late twentieth century, competition for water from urban areas has become increasingly apparent. In 1997, the then U.S. Secretary of the Interior signed regulations that permitted the first ever transfer of water from farmers to urban users. At the same time he warned California that its allocation of water was under threat because of increasing demand from Nevada and New Mexico. Population growth in Arizona, for example, will see a rise from 3.7 million to 6.4 million in the period between 1990 and 2015.[13] Groundwater over pumping has become extremely significant in the region and particularly in Arizona because of this rapidly growing population. Groundwater extraction at rates greater than what is defined as a sustainable yield has caused the groundwater levels to shrink, interfering with normal groundwater surface water interactions, and may lead to a risk of saline water incursion in some aquifers.

In 2003, the Colorado River Water Delivery Agreement between California and the basin states marked a number of proposed steps aimed at dealing with the river's over allocation and use. The agreement contained the following proposed outcomes:

- California will honor the commitment it made in 1929 by adopting specific, incremental steps to gradually reduce its use of Colorado River water over the next 14 years to its basic annual allotment of 4.4 million acre-feet (5430 GL).

- For the other basin states, Arizona, Colorado, Nevada, New Mexico, Utah, and Wyoming, the agreement provides certainty, allowing them to protect their authorized allocations and meet their future water needs.
- In the Lower Basin, Nevada, which lost access to surplus water from the Colorado River along with California, will again have access to surplus water. Nevada will be able to return to the long-term path it has developed to meet the needs of its growing population.
- The agreement enables California to provide water for its growing cities, as well as its farming communities, and to address the environmental concerns of the Salton Sea.
- The agreement allows farming communities in Southern California to strengthen their economies through water efficiency projects, canal modernization, conservation, and water marketing. In addition, divisive litigation regarding the use of Colorado River water has been resolved.
- This agreement provides the critical water supply necessary to finally resolve the water rights claims of the La Jolla, Pala, Pauma, Rincon, and San Pasqual Bands of Mission Indians.

In December 2007, a set of interim guidelines on how to allocate Colorado River water in the event of shortages was signed by the Secretary of the Interior.[14] Three levels of shortage conditions are specified depending on the level of Lake Mead (Table 2.3). However, California receives its full share of 4.4 million acre-feet (5430 GL) irrespective of declining lake levels.

Table 2.3 *Allocations to Lower Basin States Based on Lake Mead Water Levels*

Lake Mead level (AMSL)	Water Allocation (million acre-feet)
1075–1050 feet	7.167
1050–1025	7.083
<1025	7.0

As the water of the Colorado has been increasingly used for agriculture, the quantity of available water flowing to the ocean has reduced to a trickle, and the salinity of the river water has increased. The Colorado's upper reaches have a salinity of 50 parts per million (ppm). Where the river crossed the border into Mexico in the early 1900s, it was about 400 ppm. Salinity levels of water entering Mexico subsequently rose to about 1500 ppm, with levels going as high as 2700 ppm in late 1961. A dispute with Mexico resulted in the U.S. Congress passing the Colorado River Basin Salinity Control Act to deal with the problem. This act stipulated that salinity levels should be no greater than 115 ppm higher than river water arriving at the Imperial Dam, and the act included a 10,000-acre reduction in irrigable acreage in the Wellton-Mohawk Irrigation District either by purchase or eminent domain, and authorized the construction of a desalting plant at Yuma, Arizona.

As is the case in Australia, it has become evident in recent years in the Southwest United States that existing water policies, and current regulatory and management regimes that have allowed a flourishing agriculture and economy in the Southwest, are no longer appropriate to the challenges facing water demand in the twenty-first century.[15,16] Only time will tell, whether the limited institutional and policy steps described above will be sufficient to deal with the challenges caused by increasing water demand and potentially lower supply in the Colorado Basin.

Lessons from History

The four case studies presented demonstrate a number of common factors with respect to both the drivers of change in water use and the stress that water resources are facing. In all four examples, population growth and the demand for water from agriculture and the pressure that these exert on land and water resources are by far the most significant factors that have driven on ground and policy responses. As in Australia, drought and increasing understanding and concern for the environment often trigger the need for water reform. However, there are significant differences between the responses seen in the United States and Australia from those in Jordan and South Asia. In the Murray-Darling Basin, the principal response to water scarcity and stress has been the development of a system of water planning, regulation, and legally defined water rights and allocations made to users based on availability of water in storages. This has enabled water trading and a water market to evolve and has encouraged water to move from low value to high value uses. Public concern has also motivated the Australian Federal Government to step in and take over the management of the basin and implement an ambitious program to improve system efficiency and recover water saved by irrigators, for the environment. The jury is still out on whether the development of a market and having an environmental manager able to buy and sell in the market will be successful given the ongoing severe droughts experienced virtually from the beginning of the current century in Southeastern Australia. However, the fact that the market has been developed and the government has been able to buy back water to protect key environmental assets has to be commended, and it is much cheaper than major investment in infrastructure. While investment in infrastructure renewal will undoubtedly be required in some places, there is also scientific concern that water savings from minimizing leakage from canals may be unreal, in the sense that when it was leaking into groundwater, it may well have been available for use elsewhere. The issue of water rights, markets, and trading are discussed in more detail in a later chapter.

In the Southwest United States, the management of the Colorado River has seen a similar pattern of somewhat unfettered initial allocations and subsequent overuse that has had to be reined back through agreements among the states involved. However, while water is being moved from agriculture to urban uses, future population growth and climate change will bring the entire system under further stress. The signatories to the Colorado River Agreements and compacts will have to significantly review the way the waters are managed and distributed to avoid the "Armageddon" mentioned by Gertner in *The New York Times*.

There are many similarities between the examples from the Middle East and the Colorado Basin. The similarly unfettered growth in water use, consciously aided by governments and international development banks, has seen King Parakramabahu's maxim met, with little if any water (and very polluted water at that) flowing into the Dead Sea. Now, with no more water resources to turn to, Jordan is contemplating vastly expensive engineering schemes that will draw saline water from the Red Sea and use it for urban and other uses following desalination. The critical issue here is whether such an expensive scheme could be avoided or at least averted for many decades by less profligate use of the region's waters in agricultural development schemes, some of which have been seen to be of uncertain overall national benefit given the environmental and infrastructural investment consequences. Similarly, the option of importing virtual water in bulk food also has to be given close scrutiny in such water-scarce semiarid environments, thus saving local water for higher value uses.

As Tushaar Shah points out, South Asia demonstrates a different kind of response to water resources management. The region is water-scarce in some parts but not in others. However, across the region we have witnessed an explosion in the amount of area irrigated by pumps using groundwater predominantly (but not exclusively). This change seems to have occurred because of a failing on the supply side in terms of reliability and timing of water delivery to farmers, who have then, opportunistically, moved to a newly available technology and changed the physical face of irrigation systems. And perhaps more importantly, governments' failures in managing and effectively regulating water resource have had significant impact. This is putting the livelihood of millions of at risk, given that groundwater extraction in many regions, particularly in the hard rock regions of Central and Southern India, is at an unsustainable rate. Ironically, the response to water scarcity in these regions is a proposed river linking scheme that will bring water from supposedly under-utilized basins to the over-utilized basins. Again, as in Jordan, the cost of these schemes is very high, and so far there have not been adequate analyses of the social, economic, and environmental costs and benefits of what in an engineering sense is considered feasible. As will be shown later, highly expensive major river diversion schemes are schemes of last resort that should only be turned to when all other avenues have been exhausted.

The issues discussed in this chapter regarding the increase of water scarcity are generally based on concerns inherent in population growth and its resulting pressures, including the need for increased food production and urban water supplies, which greatly impact water resources and availability, as well as for the environment. Climate change will apparently also decrease water supplies in already water-scarce areas, as is suggested by the U.S. and Australian examples. As will be shown later, the pressure from consumers desiring diets richer in animal products is also starting to have a major impact on water consumption in many South and East Asian countries. On top of this, competition for land and water will come from biofuels production as well. So as is explained in the next chapter, there are a number of key drivers (population growth, dietary change, urbanization, biofuels production, environmental sustainability, and climate change) that will figure heavily on how we manage our water resources over the next 50 years.

chapter three

Causes of Water Scarcity

"Water sustains all."

Thales of Miletus, 600 B.C.

Introduction

When Colin was growing up in the 1950s and '60s, there were only about three billion people on the planet. Water seemed to be an abundant resource. Few people seemed to worry about the sustainability of water supplies, and great strides were beginning to be made to clean up polluted rivers in the developed world including the Rhine and the Thames. Through the decades after the Second World War, there was much focus on resource development and the construction of large dams and irrigation schemes in many developed and developing countries. Although global carbon dioxide emissions were increasing, no one had thought about man-made climate change and its consequences, which will be severe for some of our water resources.

By the 1990s when the world's population had reached about five billion, some tell-tale signs were emerging that indicated water was not the infinite resource that many had imagined it to be. The waters of the Colorado River in the Southwestern United States and Mexico were already under extreme pressure from the large agricultural systems of the Imperial Valley, and in Australia the Murray-Darling Basin waters were fully allocated with little left for the environment. Elsewhere, we saw the Aral Sea being desiccated by unsustainable cotton production in the then USSR. By the turn of the millennium, the Murray-Darling was intermittently ceasing to flow into the sea.

Now its channel is only kept open by dredging, and long periods of drought have severely impacted its agriculture that used to provide 40% of Australia's agricultural production. In the Middle East, Israel and Jordan face the prospect of the Dead Sea drying up because of upstream use of virtually all the water that used to flow down the Jordan River. China is planning to divert water from the Yangtze to the north China Plain, and India is contemplating a major and extremely costly river linking program to move water from rivers with abundant supply to rivers that are overallocated and overused. Groundwater resources in the United States, India, and China have similarly been overexploited and in some cases are near exhaustion.

By 2008, in spite of some regions of the world still having abundant fresh water resources, the previous examples of overuse and scarcity, coupled with a combination of regional droughts and the effects of some countries' food commodity trading policies, contributed to the development of the world food crisis. Prior to the crisis, the numbers of people living in poverty around the globe were decreasing. The crisis, which was a classic economic crisis based on price increases due to diminishing supply rather than to absolute food shortage, saw the number of the world's population living below the poverty line increase. Prior to the crisis, the Food and Agricultural Organization of the United Nations (FAO) figures indicated that about 1.4 billion people were living below the line. Similarly, those in extreme poverty (incomes less than $1 per day) increased from 850 million to over 1 billion. Many more suffered and still do so from intermittent food shortages. Living below the poverty line means that you simply do not have enough food to sustain daily activities and health. A basic premise of this book is that unless we learn to manage our water resources better and in a more sustainable manner, food crises will become increasingly frequent over the forthcoming decades and will see increasing numbers of people pushed below the poverty line, increasing malnutrition and disease and in the worst cases, growing numbers of people starving to death.

Currently the world population has reached nearly seven billion, and this large number is forecast to increase to about nine billion by 2050. Given the number of drivers and uncertainties that impinge on food production, the fact that close to one billion people are undernourished and the fact that we have already come perilously close to a famine-based disaster in 2007–2008, there has to be concern as to whether we can feed nine billion people in the future. Furthermore, as will be shown subsequently, changes in the dietary preferences in the rapidly developing countries of China and India, are putting even further demand on the provision of more water for agriculture.

Availability of enough fresh water for rainfed and irrigated cropping is probably going to be the single most important factor in our quest to feed so many more mouths. The Comprehensive Assessment of Water Management in Agriculture[1] has already indicated that if we proceed with a "business as usual" approach, water scarcity will cause us to fail in this quest. "Business as usual" means maintaining water productivity levels at their current low rates. The Comprehensive Assessment was, however, more optimistic about the future if we can substantially lift water productivity levels via a combination of biophysical and engineering inputs and policy levers. While some informed commentators indicate that we will only be able to produce enough food through expanding irrigation, the fact is that approximately 75% of production comes from rainfed areas. This is paralleled by the statistic that 80% of water evapotranspirated by crops comes from rainfed farming areas and the remaining 20% from irrigation areas.[2] However, the figures for gross value of production indicate that only 55% comes from rainfed areas, indicating the importance of irrigation in terms of its higher productivity per unit area.

This chapter examines some of the causes behind water scarcity and examines ways in which water productivity may be increased to tackle the major challenge of increasing scarcity.

Blue and Green Water

Water passing through the hydrological cycle can be viewed as "blue" or "green" (see Figure 3.1). Blue water is that found in rivers, lakes, and groundwater (aquifers). Green water can be defined as that very significant proportion of water stored in the soil. There is a complicated interaction between green and blue water, which depends on the factors (climate, topography, soil type, vegetation cover) and processes that control runoff to rivers and deep drainage to groundwater. It is the remaining green water that is available for evapotranspiration back to the atmosphere, which is the critical water supply for rainfed cropping and pasture production.

Causes of Water Scarcity

The food crisis in 2007–2008 was driven by a number of factors including increasing demand due to population growth, increasing biofuel production at the expense of food crops, regional impact of drought (such as in Australia) on agricultural water availability and thus food production, and changes in economic and trade policy in some countries.

If we look into the future with respect to water availability, not only continuing population growth, but also other factors including dietary change to more water-thirsty foods, urbanization, increasing biofuel development, and power production (thermal and hydropower) will all lead to further pressure on water resources and the environment.

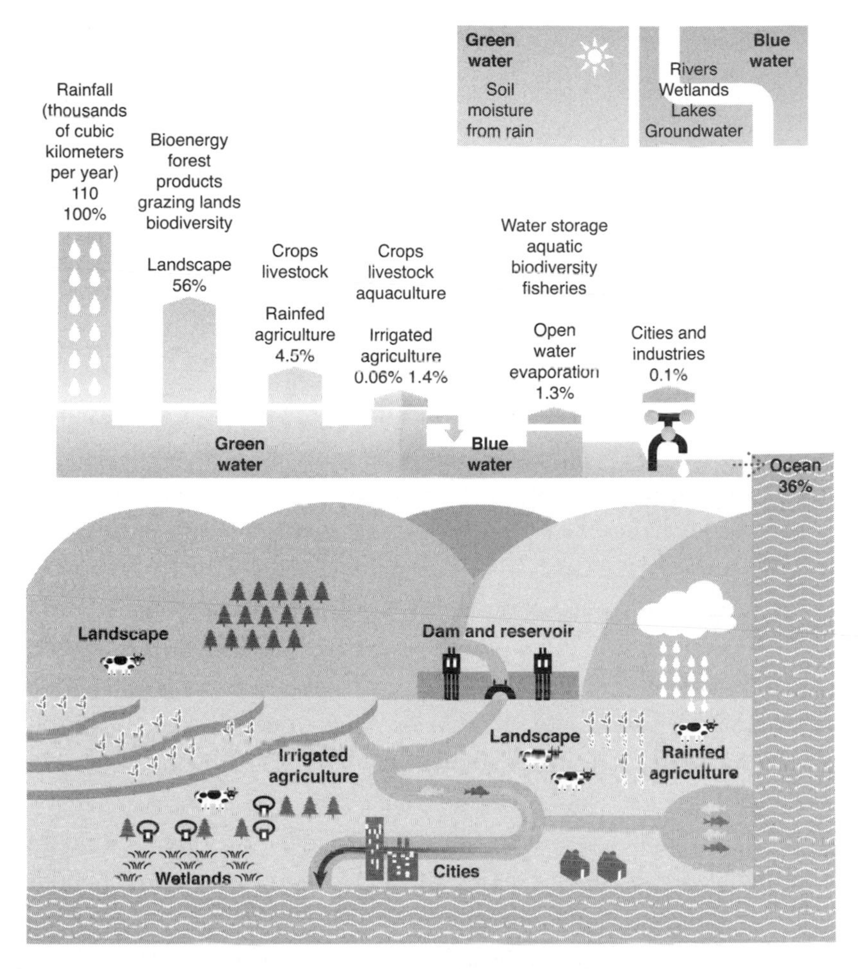

Figure 3.1 Blue and green water in the hydrological cycle

Source: Comprehensive Assessment of Water Management in Agriculture, 2007

Population Growth Impact

Global population is expected to climb from approximately 6.7 billion in 2009 to 9 billion in 2050. Given that fresh water supplies are finite, this means smaller shares for all and more widespread physical water scarcity and water stress. These stresses have symptomatic expression of scarcity, leading to potential conflict (usually political rather than physical) at a transboundary level, competition between water users, crop failure, and so on. In extreme cases physical scarcity coupled with drought leads to regional famine.

There are a number of different measures of water scarcity. The UN suggests that people need 13 gallons (50 liters) of water per day for their basic drinking, cooking, and cleaning requirements. However, such an estimate is misleading as it is a subsistence figure and does not take into account the amount people need for growing food, feed and fiber crops, raising livestock, or industrial and environmental needs. Taking these other uses into account, the FAO has used a benchmark of 1000 m^3 (264 gallons) per person per year being the level at which people face severe constraints to socio-economic development and the environment.[3] Levels of 1000–2000 m^3 (264–528 gallons) per capita can be viewed as a serious constraint and may be particularly problematical in times of drought. The FAO estimates that by 2025, 1.8 billion people will be living in countries or regions with absolute water scarcity, and two-thirds of the world's population could exist under water stress conditions.[4] The Comprehensive Assessment of Water Management in Agriculture estimates that 2.8 billion people live in areas facing overall (this includes economic and physical scarcity and as explained earlier in Chapter 1, "Not Another Crisis!") water scarcity, and more than 1.2 billion live in areas facing physical water scarcity.[5] This latter figure is almost one-fifth of the world's current population.

With respect to blue water, studies often talk of the use-to-availability ratio (so-called *criticality ratio*) and *chronic water shortage* (level of water crowding).[6] Already 1.4 billion people are living in areas where water is over-appropriated, and 1.1 billion of these people live in areas that are also suffering under severe water shortage.[7] Forecasts depict that by 2050, depending on the rate of fertility decline, the population in countries with chronic water shortages (above 1000 people per million cubic meters per year) will be 3 to 5.5 billion. To date, it has been the environment that has suffered as water use to water availability ratios rise. While many ecologists would argue that a 40% ratio is a threshold above which ecosystem health is impacted, there are a growing number of major rivers that hardly reach the sea any longer, leading to the concept of *closed basins*. Included in this number for, example, are the Murray (Australia), Yellow (China), Krishna (India), and Colorado (United States).

From a health perspective, the World Health Organization suggests that 4 out of every 10 people currently suffer from water scarcity and the health consequences of water scarcity, such as diarrheal diseases including cholera, typhoid fever, salmonellosis, other gastrointestinal viruses, and dysentery.

While these estimates of the numbers of people affected by water scarcity vary, they are essentially of the same order of magnitude. Already the majority of the world's poor live in developing countries that are affected by water scarcity to a greater or lesser degree. While population growth rates vary considerably across Asia, Africa, and Latin America, any growth in already water-scarce countries is going to exert greater stress on the existing water resources. Given that population growth is minimal, or actually negative, in many developed countries and that the majority of so-called "north" countries are not water-scarce, it is clear that population growth therefore represents the biggest single threat to water supplies and food production over much of the developing world.

Dietary Change

Previously, undernourishment because of a lack of food and appropriate dietary balance has been of primary concern to development. Indeed, there are still approximately one billion poor in the world living below the poverty threshold of $1.25 per day. This number is considered to have increased recently as a result of the 2007–2008 food crisis and its higher prices, which impact most severely on the very poor. However, as countries move up the development staircase and people's incomes rise, there is not only an increase in overall consumption, but also a shift in consumption patterns from diets rich in cereals toward those with a higher proportion of livestock products, fish, and high-value crops.

On average, to grow 1 calorie of food requires 2.1 pints (1 liter) of evapotranspirated water. For an individual on a diet without meat, about 528 gallons (2000 liters) of water per day is required to produce it compared with 1320 gallons (5000 liters) per day for a diet high in grain-fed beef.[8] However, if meat is produced from rangeland or rain-fed pastures, its water consumption is likely to be considerably less than for grain-fed cattle. Data from the FAO show very considerable variation between diets in different regions of the world.[9] For example, meat consumption in a basket of industrial countries averaged approximately 200 pounds (90 kg) per person per year in 2001. In comparison consumption in sub-Saharan Africa and East Asia was 88 pounds (40 kg) and 13 pounds (6 kg), respectively.

If we require an average daily calorie intake of 3000 (which allows for some production losses and wastage), then to feed the predicted 2050 world population of 9 billion, we will need to find approximately an additional 600–720 cubic miles (2500 to 3000 km^3) of fresh water. This figure is probably conservative as it is based on a relatively low protein diet and minimal food waste post farm gate.

Figures drawn up for the Water Education Foundation of California shows that a "typical home serving" of food required the following water volumes to produce it: 2.9 gallons per 1 cup of lettuce (11 L per cup), 7.3 gallons (28 L) per slice of white bread, and 1232 gallons (4660 L) per 8 oz. (0.23 kg) of steak.[10] Other reputable sources such as UNESCO quote the requirement for over 30 gallons (140 L) of water to produce 1 cup of coffee. However, it is important to consider that many of these foods and beverages contribute to the quality of daily life. What, therefore, is important is not that water is used in their production, but that the use is as efficient and productive as possible in order to minimize total water consumption.

Biofuel Production

The global financial crisis has seen oil prices tumble from US$150 per barrel to $37–40 and then back to about $70–80 during 2008 to 2010. While the crude oil price will undoubtedly have a short-term impact on the attractiveness of ethanol production, growing fossil fuel demand over the forthcoming decades will inevitably result in an increase in demand for biofuel as well. So-called first generation biofuel production derived from corn, soybeans, and sugarcane create competition for not only land, but also water. India and China currently rely on irrigation (blue water) to produce these crops, whereas Brazil and the United States, whose ethanol industries are based on sugarcane and corn, respectively, rely predominately on rainfed agriculture (green water).

Currently many countries are considering policies that stipulate a biofuel production target. Even where these are as low as 10–15% they will have significant impact in terms of competition for land and water resources as well as potentially increasing the source crop's price.

Growing sugarcane to produce the 9 billion liters of bioethanol needed to meet just 10% of India's gasoline fuel demand by 2030 could require another 22,000 billion liters of irrigation water. And this is assuming that water efficiency improves. By 2030, India's demand for cereal is set to rise by 60%—and more than double for sugar. Analysis by the International Water Management Institute (IWMI) indicates that even under the most optimistic scenario, the demand for irrigation water will increase by 13%—equivalent to 84,000 billion liters, or roughly the annual flow of the Krishna River in Andhra Pradesh, India. If demand for biofuel shifts water supplies away from food production, then the national government must be forewarned about the impact it could have on food supplies so that alternative measures, including importing more staple foods, can be put into place.

According to an article in the *New Scientist*, Canada would have to use 36% of its farmland to produce enough biofuel to replace 10% of the fuel currently used for transportation. Brazil, by contrast, would need to use only 3% of its agricultural land to attain the same result.[11] So what might make sense in one region or country might not make sense elsewhere.

If in this competition, the biofuel producers ultimately take over 10–15% of agricultural land, the impact on food production will be very significant. Where biofuel production requires irrigation in physically water-scarce countries, it will be at the expense of water for crops.

There are, however, many uncertainties on the future impact of biofuel on food production: biofuels' rates of return; concern that their production is no more greenhouse-friendly than fossil fuels; and the potential advantages of new technologies that will enable production of biofuel from algae, crop residues, and other wastes (so-called second-generation biofuel).

Urbanization, Globalization, and Other Factors

In 2008, we saw an overall transition from a world in which a majority of people lived in rural environments to one with more people in the cities and towns. While the turnover point has not yet been reached in the developing world, it is likely to be inevitable. Bigger cities with more industry clearly already compete with agriculture for water resources, and this competition will continue to increase. Furthermore, cities generally have political power and the wealth to purchase water from other users. Currently, many agricultural developing countries and developed countries/states such as Australia and California utilize 70% or more of their total available water resources in the agricultural sector. Even if growing urban demand only requires a redistribution of 5% of this water, this will have a significant impact on agricultural production from a global perspective.

Globalization of food supply chains dominated by major supermarket companies will also have a range of impacts on food production. These will include demand for luxury goods such as cut flowers creating competition for land and water in regions close to international airports, product sourcing policies of supermarket chains in developing countries, and the recent phenomenon of large food-importing countries wanting to buy up large tracts of land in developing countries for food production. This so-called land grab is real and is happening today, and unless it is carefully considered and regulated, it will have unhappy consequences for land-rich countries and their poorer communities. Competition for water from the hydropower industry can also mean that water for agriculture will no longer be available at the right place at the right time. Additionally, much of the water used for cooling thermal power plants is evaporated and lost. These types of drivers may significantly impact food production in developing countries and have both beneficial and adverse impact on their populations. Lastly, the negative impact of restrictive world trade policies cannot be overlooked as a factor that may, in countries with abundant water and food production potential, limit the development of their production and marketing capabilities and thus their capacity to contribute to world food production as times get tougher.

While domestic water requirements are relatively low, increasing pressure to meet the Millennium Development Goals in terms of water supply and sanitation for the poor does cause competition for both the water resource and scarce financial resources in the water sector.

Finally, ongoing loss of agricultural land due to urbanization and industrial development and land and water degradation (often the result of population pressure) will continue to destroy or reduce the productivity of agricultural land.

Climate Change Impact

The impact of climate change on water resources will be different from place to place. The results of the Intergovernmental Panel on Climate Change (IPCC) report on climate change and hydrology are worrying to say the least.[12] While there is not enough large-scale data to facilitate prediction at scales appropriate to agricultural water management as yet, some issues are of major concern and present critical threats to water availability and food security. These include reduced snow melt inputs into many significant river systems and the likelihood that subtropical/lower mid-latitude environments may suffer most from temperature rise and rainfall decreases. Chapter 4, "Climate Change and Water," provides more information on the likely trends and the impact of climate change on agriculture.

Conclusion

While the trends described in this chapter are worrying with respect to their impact on water resources, there are many ways in which wealthy and less wealthy communities and individuals can adapt to them. These include improving water storage, improving water productivity (more crop per drop), demand management processes, trading in virtual water, and other innovations. These and other potential strategies for coping in an increasingly water-scarce world are presented later in the book.

chapter four

Climate Change and Water

"All earth's full rivers cannot fill
The sea that drinking thirsteth still."

Christina Rossetti (1830–1894), *By the Sea*

Of all the factors affecting future water supplies and agriculture and food production, climate change is the one that we understand least in terms of the likely magnitude, severity of impact, and geographic expression and variation. This is not surprising because not only is much of the science of climate change new, but it has also only been in the last few years that the army of climate change skeptics have begun to accept what was originally a hypothesis as now becoming a reality. The 15 years of bickering over whether or not climate change is caused by humankind has meant much time and effort that could have been devoted to mitigation and adaptation has been lost in debate about the causes. In terms of agriculture and water resources, the cause is relatively unimportant compared to deciding how we can deal with the impact.

However, while uncertainty prevails over the impacts of climate change, it is clear that investment in adaptation technologies is money well spent because these will also combat the other causes of water scarcity and food insecurity.

In the post Kyoto Climate Change Conference of 1997 environment, there has been increasing discussion about how agriculture and land use factors can be included in overall assessments of greenhouse gas sources and sinks. Agriculture contributes about 14% of annual greenhouse gases.[1] However, land clearing and the development of wetlands for agriculture significantly adds to this by another 19%. Fifty percent of agricultural emissions come from the developing world, which also contributes 80% of land clearing-related emissions. When agriculture is considered in the context of the entire food supply chain, estimates suggest that it accounts for about 20–22% of emissions. As has been demonstrated previously, agriculture entrains the largest share of the world's extracted water resources, so there is a strong if somewhat indirect link between water use and climate change. Furthermore, where energy is used to pump and deliver water, this can be a direct producer of carbon dioxide emissions unless the energy is generated by hydro or nuclear plants. Consequently, it is very important that water development projects consider not only whether they meet the highest standards of efficiency and water productivity, but also their impact on carbon dioxide emissions.

In terms of the relationship between climate change mitigation and agriculture and water, improved tillage practices can sequester carbon—although results seem to vary considerably with geography and climate. Similarly, potential higher crop productivity can mean less land required globally for crops, thus reducing the rate and amount of deforestation. Increasing the use of hydropower also reduces the need to burn fossil fuels to generate electricity, and clearing timber from reservoir sites prior to inundation can help to reduce methane emissions.

However, the impact that climate change will have on water resources and food production is the chief concern of this book. Consequently, this chapter examines some of the most probable impacts of climate change on water resources and food production systems and examines some adaptive management strategies that will be required if impacts are to be minimized.

Preliminary analysis of climate change forecasts does, however, suggest that many developing countries will bear the brunt of its impact on water resources and agriculture. While it can be argued that warmer and wetter conditions in higher latitude countries may bring more land into production to compensate for production losses in lower latitudes, this does little to assist poor farmers and their families bearing the impact of increasing droughts, flooding, and higher temperatures. Many developing countries do not have the foreign exchange to import food from developed countries, nor do they want to contemplate the major social disruption that would come from many rural poor losing income and livelihoods. Such outcome would lead to famine, social unrest, illegal migration, more people smuggling, and potentially terrorism. So the development of sensible, low-cost adaptive management strategies to deal with climate change in developing countries has to be an international imperative.

What Are the Most Likely Impacts of Climate Change?

> "Observational records and climate projections provide abundant evidence that freshwater resources are vulnerable and have the potential to be strongly impacted by climate change, with wide-ranging consequences for human societies and ecosystems."[2]

This statement prefaces the Executive Summary of the International Panel on Climate Change's report on Climate Change and Water. The key conclusions of the Intergovernmental Panel on Climate Change (IPCC) on water resources are shown in the next sidebar.

While there have been some studies that suggest that higher CO_2 levels could, in the presence of adequate water, boost crop yields and tree growth, it would appear from the IPCC's work that any such effects in the lower-mid latitudes and subtropics will be offset by lower rainfall and higher temperatures that cause more evaporation. Their work also suggests that increased warming due to anthropogenic influences causes more water vapor to accumulate in the troposphere and may account for the increasing frequency of very wet days that has been observed from a number of locations across the globe. Similarly, they suggest an increasing degree of variability with more cyclones and more droughts. There is also clear evidence of decreasing snow cover in mid latitudes and significant mass loss on glaciers and icecaps worldwide.

The IPCC report indicates that at the global scale, there is evidence of a broadly coherent pattern of change in annual runoff, with some regions experiencing an increase in runoff (for example, high latitudes and large parts of the United States) and others (such as parts of West Africa, southern Europe, and southernmost South America) experiencing a decrease in runoff.[3] However, it is notoriously difficult to separate climatically induced changes in runoff from those caused by other factors, such as land use change.

From the point of view of water scarcity and availability, the predicted decrease in runoff in subtropical and lower-mid latitude regions and potentially shorter and more severe monsoons are probably the cause for greatest concern in the immediate future, and the decline in runoff from snowmelt and glaciers is the greatest over a slightly longer period.

Key Summary Points from the IPCC Report on Climate Change and Water Resources

- Observed warming over several decades has been linked to changes in the large-scale hydrological cycle.
- Climate model simulations for the 21st century are consistent in projecting very likely precipitation increases in high latitudes and parts of the tropics, and likely decreases in some subtropical and lower mid-latitude regions.
- By the middle of the 21st century, annual average river runoff and water availability are projected to increase as a result of climate change at high latitudes and in some wet tropical areas, and decrease over some dry regions at mid-latitudes and in the dry tropics.

- Increased precipitation intensity and variability are projected to increase the risks of flooding and drought in many areas.
- Water supplies stored in glaciers and snow cover are projected to decline in the course of the century.
- Higher water temperatures and changes in extremes, including floods and droughts, are projected to affect water quality and exacerbate many forms of water pollution.
- Globally, the negative impacts of future climate change on freshwater systems are expected to outweigh the benefits.
- Changes in water quantity and quality due to climate change are expected to affect food availability, stability, access and utilization.
- Current water management practices may not be robust enough to cope with the impacts of climate change.
- Climate change challenges the traditional assumption that past hydrological experience provides a good guide to future conditions.
- Adaptation options designed to ensure water supply during average and drought conditions require integrated demand-side as well as supply-side strategies.
- Mitigation measures can reduce the magnitude of impacts of global warming on water resources, in turn reducing adaptation needs.
- Water resources management clearly impacts many other policy areas.
- Several gaps in knowledge exist in terms of observations and research needs related to climate change and water.

Source: Bates et al., 2008 (see endnote 2).

Heavier rainfall causing erosion of agricultural land and flooding will also be of increasing concern. In coastal environments, the loss of crop land to seawater inundation as sea levels rise and saline groundwater incursion are already becoming problems and will get steadily worse through the century.

While different climate models show convergence with respect to lower precipitation in the subtropics and lower mid-latitudes, there are some differences. For example, the two scenarios in Figure 4.1 respectively show South Asia drier and South East Asia wetter. Mean changes for the period 2080–2099 relative to 1980–1999 from up to 15 models are shown in Figure 4.2.

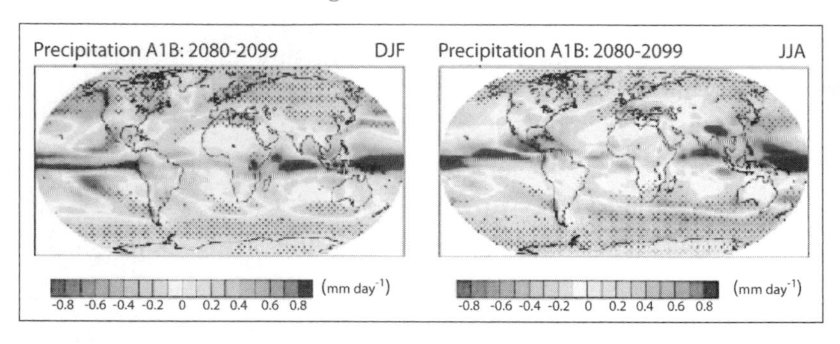

Figure 4.1 IPCC forecasts for different models for global precipitation in the period 2080–2099

Source: IPCC 2008, *Climate Change and Water*. Technical Paper of the Intergovernmental Panel on Climate Change (IPCC), Figures 2.7 and 2.8. IPCC Secretariat, Geneva

Significant problems, however, are involved in relating the results of global climate models to specific changes in rainfall, evaporation, and runoff at catchment level. These are due to the inherent uncertainties involved with the models themselves and the assumptions made as well as the difficulties involved in downscaling global estimates to local regions. Currently, considerable effort is being put into these downscaled modeling exercises. When complete, their results will provide much improved data, which can then be used in basin and catchment hydrological modeling exercises.

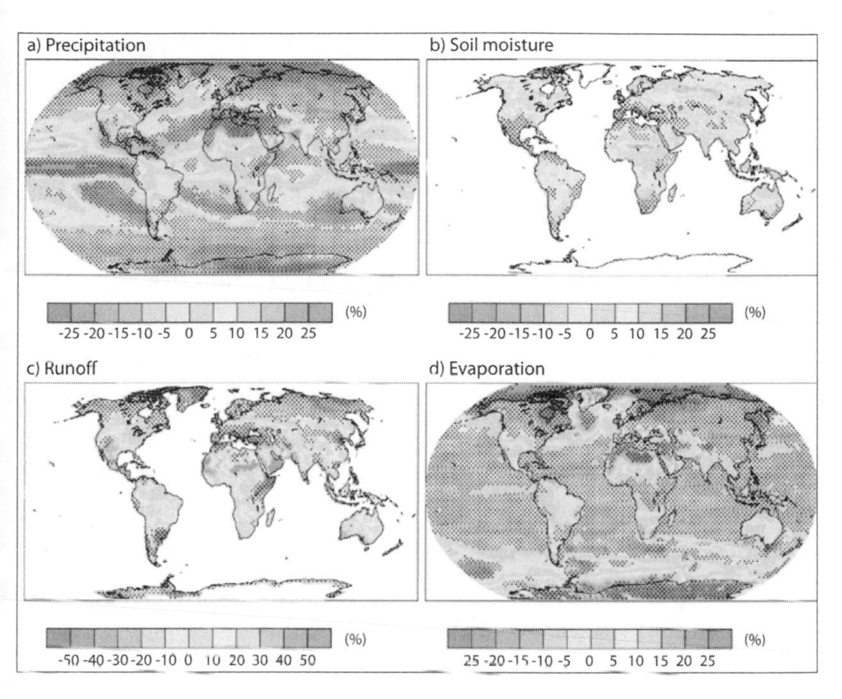

Figure 4.2 Fifteen-model mean changes in (a) precipitation (%), (b) soil moisture content (%), (c) runoff (%), and (d) evaporation (%). To indicate consistency of sign of change, regions are stippled where at least 80% of models agree on the sign of the mean change. Changes are annual means for the scenario SRES A1B for the period 2000–2099 relative to 1980–1999. Soil moisture and runoff changes are shown at land points with valid data from at least ten models.

Source: IPCC 2008. *Climate Change and Water.* Technical Paper of the Intergovernmental Panel on Climate Change, Figures 2.7 and 2.8. IPCC Secretariat, Geneva

While many people are skeptical of modeling exercises, there is a building body of actual measurement that is starting to confirm the predictions. Furthermore, there are clear instances from the relatively recent geological past that demonstrate that climate change (induced by other than anthropogenic processes) is not as rare as we might imagine. Examples include the much colder weather experienced during Europe's Little Ice Age in the eighteenth century and the Younger Dryas period of about 12,000–13,000 years ago, which saw extreme cooling that lasted for about 1000 years. Forecasts made for the Australian rangelands[4] suggested that climate change is likely to reduce rainfall in the rangelands, which could lead to a 15% drop in grass productivity. This, in turn, could lead to reductions in the average weight of cattle by 12%, significantly reducing beef supply. Under such conditions, dairy cows are projected to produce 30% less milk, and new pests are likely to spread in fruit-growing areas. Additionally, such conditions are projected to lead to 10% less water for drinking. Studies conducted for the Pentagon[5] suggested that if similar conditions as described in Australia occur in several food-producing regions around the world at the same time within the next 15–30 years, there could be widespread famine and environmental disasters, challenging the notion that society's ability to adapt will make climate change manageable. The authors of this report stressed, however, that the effects anticipated were not a prediction of imminent doom, but a scenario of the worst possible outcomes of abrupt climate change that could cause a social and economic catastrophe. In 2007–2008, the world did see a food crisis that was caused by a combination of drought, supply and demand imbalances, and government policies diverting water and land to biofuels production. This suggested that the scenarios painted by futurologists can have some credence, and it is important to start planning for difficult conditions ahead indicated by scenario analyses.

Climate Change Impacts in Asia

A report in the *Pakistan Daily Times* of June 5, 2009, pointed out ominous outcomes of climate change for Pakistan. It reads:

> "Indian defense scientists have found that temperature in the North-Western Himalayan range, spread over Jammu and Kashmir and Himachal Pradesh and the source of most of Indus line rivers and their tributaries, have risen by nearly 1.4

> degrees Celsius over the last 100 years. The average level of warming for the rest of the globe is 0.5 to 1.1 degrees Celsius per century. They also found that lesser snowfall, followed by an early thawing of snow, has resulted in a changed water balance in the catchments. Since Pakistan gets most of its water from these catchments, the study forecasts a disastrous situation for the coming years. The study said the glaciers in the Himalayan region had shrunk considerably in the last three decades, indicating a major ice-loss."[6]

This picture is mirrored elsewhere in most mountainous areas. Mountain regions at risk include the Himalayas, Tien Shan, and the Pamirs of Tajikistan but also the Andes and the European Alps, according to a recent UN Environment Programme (UNEP) report, Global Outlook for Ice and Snow.[7] UNEP projected that an estimated 40% of the world's population, hundreds of millions of people, could be negatively affected by loss of snow and glaciers in the mountains of Asia, including reductions in snow cover, sea ice, glaciers, permafrost, and lake ice. The report also highlighted a further hydrological risk resulting from climate change—the formation of lakes as a result of melting glaciers, which increases the risks of glacial lake outburst floods. Such lakes could potentially release up to 100 million cubic meters of water down into vulnerable valleys.

In Central Asia, it has been estimated that the glaciers in Tajikistan lost a third of their area in the second half of the twentieth century alone, while Kyrgyzstan has lost over a 1000 glaciers in the last four decades. Glaciers in the region, numbering at this point 116, have lost nearly two cubic kilometers of ice a year between 1955 and 2000 due to a small but pervasive rise in temperature. Reports for a catchment (watershed) in the Tien Shan Mountains of Kyrgyzstan state that there was an overall loss of the area of glaciers of 28% for the period 1963–2000, and a clear acceleration of wastage since the 1980s correlate with the results of previous studies in other regions of the Tien Shan as well as the Alps.[8] The report further indicated that glaciers smaller than 0.5 km^2 have exhibited this phenomenon most starkly. While the report registered a medium decrease of only 9.1% for 1963–1986, they lost 41.5% of their surface area between 1986 and 2000.

The expected chain of hydrological events associated with climate change in glaciated regions has been described as consisting of increased flood risk as a first stage followed by a complex picture after a complete glacier loss.[9] This would include higher discharge during spring due to an earlier and more intense snowmelt followed by a water deficiency in hot and dry summer periods. This unfavorable seasonal redistribution of the water supply has dramatic consequences for the Central Asian lowlands, which depend to a high degree on the glacier melt water for irrigation and already now suffer from water shortages. It is anticipated, for example, that flows in the Syr Darya and Amu Darya Rivers, which drain into the Aral Sea, may decline by about 30% as glaciers disappear. This will be catastrophic for the region as inefficient water use has already seen complete allocation of the rivers' flow, which was a major cause of the drying up of the Aral Sea. Furthermore, conflicts about water use priorities also occur in this region with major demands for hydropower-based electricity supplies occurring in winter when the water discharged from power plants in Kyrgyzstan and Tajikistan cannot be used for irrigation. This presents downstream countries of Uzbekistan and Kazakhstan with winter flooding and summer water shortage at the time when irrigation requirement is at a maximum. Much more research is needed to discover whether some of this water can be restored downstream in new barrages or as groundwater and thus made available for irrigation.

Some commentators have suggested that all the glaciers in the Central Asian region will have melted away in the next 30 years, although other scientists are more conservative. A similar issue for the Himalayas gained much press attention in late 2009 to early 2010 when it was established that the IPCC had reported unsubstantiated data and assumptions about ice loss. The specific quote was taken from a World Wildlife Fund (WWF) report, "In 1999, a report by the Working Group on Himalayan Glaciology (WGHG) of the International Commission for Snow and Ice (ICSI) stated: 'Glaciers in the Himalayas are receding faster than in any other part of the world and, if the present rate continues, the livelihood[sic] of them disappearing by the year 2035 is very high.'" Quoting the respected International Center for Integrated Mountain Development (ICIMOD):

> "Many of the inferences regarding glacial melting are based on terminus fluctuation or changes in glacial area, neither of which provides precise information on ice mass or volume change. Measurements of glacial mass balance would provide direct and immediate evidence of glacier volume increase or decrease with annual resolution. But there are still no systematic measurements of glacial mass balance in the region although there are promising signs that this is changing. China is the only country in the region which has been conducting long-term mass balance studies of some glaciers and it has expressed the intention of extending these to more Himalayan glaciers in the near future. India has recently started to study several glaciers for regular mass balance measurements. Recognizing the importance of mass-balance measurements, ICIMOD has been promoting mass balance measurements of benchmark glaciers in its member countries and has co-organized trainings to build capacity for this in the region."

So the jury is still out to some extent on the rate and degree of ice loss, but it is hard to find concrete evidence that refutes the fact that this is happening given rising temperatures. Consequently, it would appear that substantial loss of ice and snow cover in many mountainous reasons is not subject to a question of *if* but *when*. The shrinking and anticipated disappearance of many of the world's glaciers has potentially catastrophic consequences for communities that rely on ice melt for water for irrigation, drinking, and hydroelectric and nuclear power stations.

Climate Change Impacts in Australia

Given the already felt impacts of drought in Southeastern Australia (see Figure 4.3), the National Water Commission funded the CSIRO in 2005 to do a major study of water resources in the Murray-Darling Basin. This study led to the world's largest basin-scale investigation of the impacts on water resources of catchment development, changing groundwater extraction, climate variability, and climate change. Its terms of reference were to estimate current and likely future (~2030) water availability in each catchment/aquifer and for the entire MDB considering:

- Climate change and other risks
- Surface-groundwater interactions

The study's findings included the following key points:

1. Water resource development has caused major changes in the flood regimes that support important lakes and wetlands.
2. The south of the MDB was in severe drought from 1997 to 2006—in places, a 1 in 300 year event without climate change. The drought has continued into 2009.
3. Under the median 2030 climate, water availability would fall by 11%–9% in the north and 13% in the south.
4. The range of possible climate outcomes is wide due to the uncertainty inherent in current climate models.

Figure 4.3 Drought and probable climate change impacts have substantially reduced the amount of water available for irrigation over the last decade in Southeastern Australia. Here, near Mildura in the State of Victoria, pumps are used to lift water for irrigation from the River Murray.

Photo: Colin Chartres

What was also of considerable interest is that the conditions experienced under the 1997–2006 drought were as extreme as those predicted by climate models for 2030. A combination of development and drought meant that the River Murray ceased to flow into the ocean for about 40% of the time compared with an estimated 15% under natural conditions, and total flow was reduced by 61%. The most salient point in the report is the fact that water availability decreased by 9–13%. Even more pronounced decreases in runoff, which appear to be related to declining rainfall, have been observed in Southwest Australia, as demonstrated in Figure 4.4. If we consider that the impacts of climate change are likely to be similar elsewhere, we need to explore just how more densely populated countries often with less resilient social and economic systems will be able to deal with similar declines. Furthermore, these declines go right to the heart of one of the key theses of this book; that is, how can we grow more food with less water?

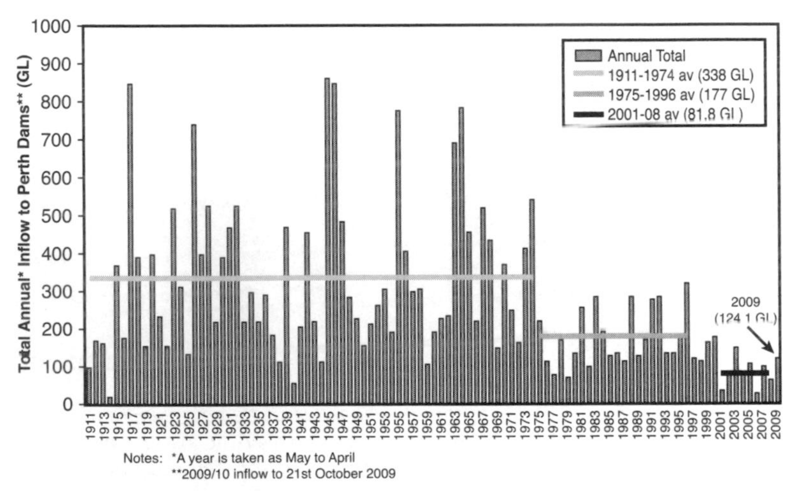

Figure 4.4 Impact of climate on water availability—reduced inflows to dams. The above data from Perth, Western Australia demonstrate major reductions in runoff into the cities' dams over the last 100 years. Whether this is due to climate change or variability is immaterial, in that the city has had to look for new sources of water including desalination.

Acknowledgments to Water Corporation, Perth, Western Australia

Does Water Have a Role in Mitigating Climate Change?

Responding to the impacts of climate change through adaptive management strategies that look at increasing water storage in reservoirs and groundwater, the development of more drought-tolerant crop cultivars, and new irrigation and supplementary irrigation schemes will be a major challenge for agriculture in general. It is true that if we can increase agricultural productivity on existing farm land and use existing water withdrawals, we can halt or minimize the spread of agriculture into forested areas, thus mitigating or delaying further greenhouse gas emissions. Similarly, encouragement of conservation agriculture that sequesters carbon can play a role in mitigating greenhouse gas emissions, as long as the carbon remains locked up in the soil. New research into locking carbon up as biochar (charcoal created by heating that differs from charcoal only in the sense that its primary use is not for fuel) also offers some prospects in this direction as well. Encouraging people to eat less meat will also reduce pressure on expanding agricultural frontiers, particularly where forests are replaced by grazing land and/or cropping to grow grain to feed animals.

Adapting to Climate Change

As mentioned at the beginning of this chapter, the type of adaptive responses required to cope with climate change are often very similar to the responses needed to adapt to other causes of water scarcity, such as increasing demand and competition. The final chapter of this book looks at a range of potential adaptive solutions. However, it is worth mentioning here that there are some specific ways of adapting to climate change with respect to ensuring food security in the developing world. These include

- Building flood and drought early warning systems
- Improve all forms of water storage including above ground and below ground (groundwater) to provide "insurance" against drought and climate change.
- Developing appropriately sized smallholder supplementary irrigation and water harvesting schemes in drier regions

- Development of large dams for irrigation and hydropower uses
- Improving rangeland resource management practices
- Managing wetlands for community use, fishery production, and conservation purposes
- Encouraging soil improvement through conservation agricultural practices to improve water holding capacity and to sequester carbon
- Encouragement of agroforestry (integrated approach of using the interactive benefits from combining trees and shrubs with crops and/or livestock) to provide fuel and minimize erosion
- Development of drought-resistant crop varieties (see Figure 4.5)
- Improved availability of crop insurance and other farmer financial support mechanisms, such as microcredit, to carry farmers through bad years

Figure 4.5 Breeding more drought-tolerant cereal varieties will be a key strategy to combat climate change in semi-arid regions. This picture shows the experimental farm at Aleppo, Syria on the site of the International Center for Agriculture in Dry Areas.

Photo: Colin Chartres

Conclusion

This chapter demonstrated that climate change presents an enormous threat to water resources, food production, and livelihoods. In our view the seriousness of this threat has not been properly appreciated as yet by politicians and policy makers or the general public, except perhaps in those areas where the impacts are beginning to bite. It is also a certainty that given our limited understanding of the impacts of climate change and the contentious nature of the debate about it that has raged for over 20 years, that there is a need for significantly more high-quality research that takes climate change scenario outcomes and uses them as inputs into hydrological modeling and planning processes, as has been started in Australia. When the best available information is on the table, a basis for a more rationale and informed debate in the community about what responses are required can be realized. However, in our opinion time is running out, and action on the impacts of climate change on water availability and food production is needed in the short term. Subsequent chapters provide more detail with which we can adapt water use and agricultural systems not only to the impacts of climate change, but to increasing water scarcity caused by increasing populations and associated domestic and industrial demand for water, dietary changes, and competition from bioenergy production.

chapter five

Agriculture and Water

"Truths are first clouds; then rain, then harvest and food."

Henry Ward Beecher

Agriculture is by far the biggest consumer of water at the global level. Agriculture is also one of the sectors of the economy in most developing countries that employs many poor people. Consequently, agriculture, water, and poverty are inexorably linked. In the developing world virtually any steps aimed at improving agricultural productivity can have beneficial or negative impact on the poor. For example, one way to increase productivity is for fewer larger farms to use water more efficiently. However, this might negatively impact the poor by reducing employment opportunities. In democratic countries like India, the relationship between agriculture, water, and poverty is clearly demonstrated by the strong clout of the rural poor when it comes to state and federal elections. Thus, reforms that many in the West and the international finance agencies see as aimed at increasing water and food productivity are often viewed with considerable suspicion by the rural poor involved in farming. For example, direct subsidies on agricultural production costs might be much better spent on rural infrastructure, irrigation development, or capacity building. However, the political reality is that many subsidies on water, electricity, and fertilizers prevail in developing countries because politicians are loathe to make reforms that may see a rural voter backlash at the next election. Because of the complex relationship between agriculture, water, and poverty, we have devoted this and the next chapter to attempting to describe both the physical dependence of agriculture on water (this chapter) and the socio-economic forces that drive agricultural water management in the context of poor countries (Chapter 6, "Water, Food, and Poverty"). However, to place all these factors in context, some overarching remarks about these three integrated elements are given here.

Water and food are inescapably linked. Many developing countries and some "western" economies including Australia and California use more than 70% of their total water extractions for agricultural production. The previous chapter's analysis of the drivers of water availability demonstrates that it will be foolhardy for any water users to overlook water scarcity issues in the future. And food crises caused by demand/supply perturbations inevitably impact the poor. Prior to the food crisis of 2007–2008, it was estimated that about 1.4 billion people were living below the poverty line of $1.25 per day. Many of these people also suffer from malnutrition. The World Bank forecasted in 2005 that the Millennium Development Goal of reducing extreme poverty 50% from its 1990 level would be realized by 2015. While this will likely be achieved, the 2007–2008 food crisis and potential future similar crises may now create some uncertainty with respect to this forecast. The ability of water policymakers and managers to ensure that agricultural water supplies are available and used more productively than at present will be fundamental to achieving the Millennium Development Goal.

Most of the world's poor live in South and East Asia and sub-Saharan Africa (see Figure 5.1), and the majority of the poor live in rural environments and are dependent on agriculture for employment (see Figure 5.2). Many of the countries in this region of Asia, but particularly Pakistan, India, and China, already have water availability concerns. India and China are described by the World Development Report as emerging economies.[1] They are transforming from agricultural to industrial societies with concomitant increases in water demand from their industrial and domestic sectors. At the same time, both countries are developing a taste for diets richer in animal products that consume more water to produce. They also have increasing populations, and neither is immune from climate change impacts on water supply and agriculture.

So India and China both exemplify the paradox of having to produce more food with less water. With limited opportunities to increase water availability and given the increasing competition for water in water-scarce agricultural countries as well as the potential impacts of climate change, it is not illogical to assume that over the next 40 to 50 years, the share of water going to agriculture will decline by perhaps as much as 10% in the face of competition from other sectors of the economy. If biofuel production using existing technologies also increases significantly in this period, as is policy in some countries, the proportion of water going to food and feed production may decline even more. Yet, as will be shown subsequently, food production targets suggest that agriculture will need significantly more water if we are to maintain global food security and feed the 2.3 billion more mouths expected by 2050.

Agricultural water management options range across a spectrum from entirely rainfed, through supplementary irrigated, to fully irrigated. All these systems have a valuable role in the food supply chain. Furthermore, all such agricultural systems interact in complex ways with natural ecosystems. In rainfed systems, crop and animal productivity varies tremendously depending upon how soils are managed structurally. Key beneficial techniques include maintaining surface vegetation cover to stop erosion and loss of valuable organic topsoil; maintaining organic residues and using minimum tillage to improve structure and water holding capacity; and on sloping lands, methods such as contour plowing, erosion banks, and other soil conservation technologies. Irrigation systems, whether supplementary or full, range from flood irrigation of small plots to large fields through small and large sprinkler systems to drip systems that can also include fertilizer

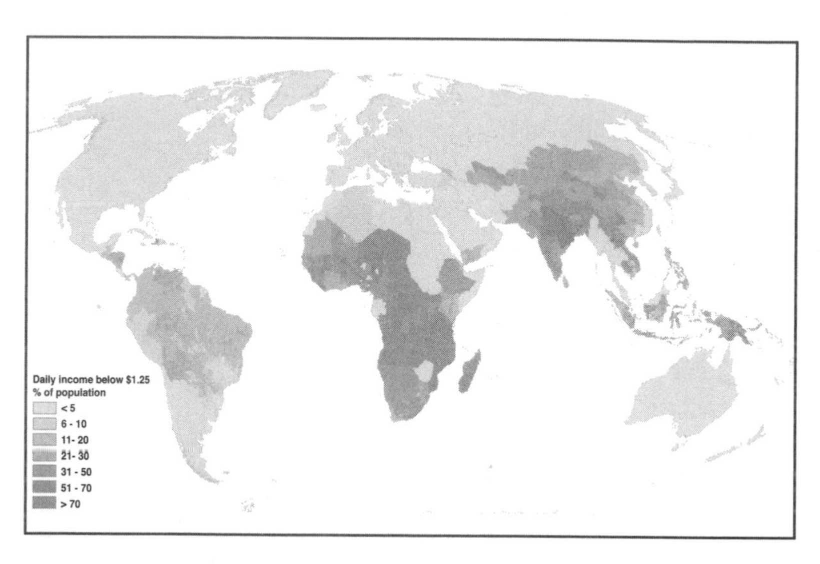

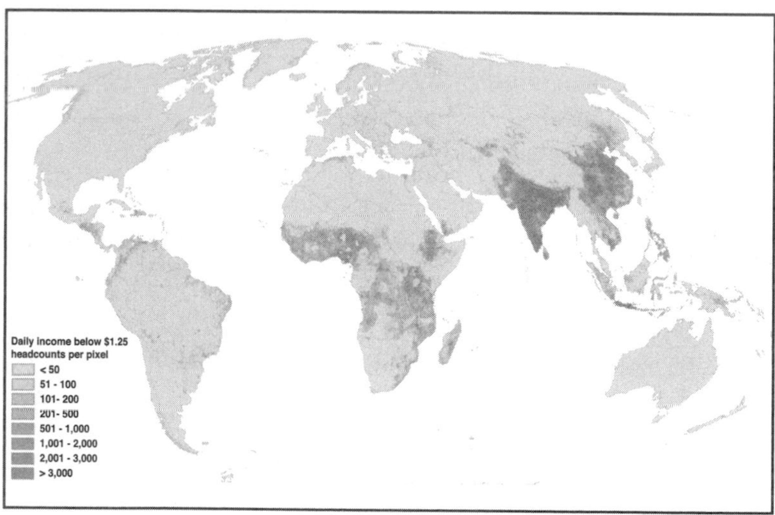

Figure 5.1 Proportions and absolute numbers of people living on less than US$1.25 per day in 2009

Source: *Geographic Domain Analysis to Support the Targeting, Prioritization & Design of a CGIAR Mega-Project (MP) Portfolio.* Draft Progress Report. September 9, 2009.

Dataset taken from the authorship of the report on the poverty mapping: Stanley Wood[a], Glenn Hyman[b], Uwe Deichmann[c], Elizabeth Barona[b], Ria Tenorio[a], Zhe Guo[a], Silvia Castano[b], Ovidio Rivera[b], Enna Diaz[b], John Alexander Marin[b].

Additional data and technical contributions from: Harold Coulombe[c] and Quentin Wodon[c], Maria Muniz[d], Sam Benin[a], Todd Benson[a], Julio Berdegue[e], Jesus Gonzalez[e], Ana Maria Barufi[e], Peter Lanjouw[c] and Ken Simler[c], Brian Blankespoor[c] and Siobhan Murray[c].

[a]International Food Policy Research Institute, Washington D.C., United States

[b]CIAT-International Center for Tropical Agriculture, Cali, Colombia

[c]World Bank, Washington D.C., United States

[d]Center for International Earth Science Information Network, Earth Institute at Columbia University, New York, United States

[e]Rimisp-Centro Latinoamericano para el Desarrollo Rural, Santiago, Chile

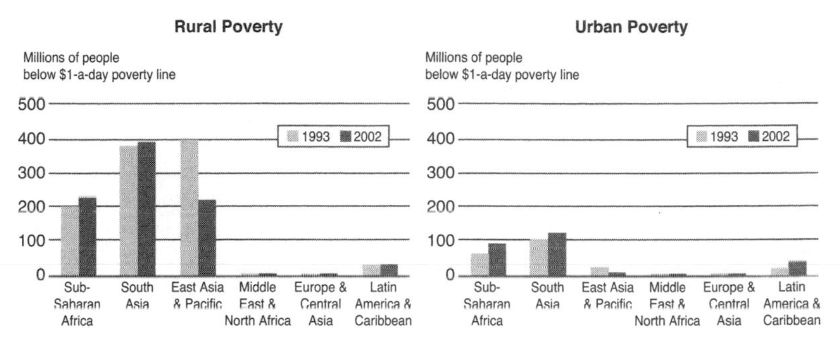

Figure 5.2 Rural and urban poverty—1993 and 2002

Source: Ravallion, et al., 2007[2]

applications in the irrigation water (see Figure 5.3). When used efficiently, all of these agricultural water management systems can be highly productive. Gravity systems generally call for more water application than pressurized sprinkler and drip systems because of uneven fields, long furrow runs that end up with too much water near the source and too little at the end of the furrow and from high evaporation losses. However, precisely designed and level fields show a high level of application efficiency (see Figure 5.4). Sprinkler and drip systems can be more precisely controlled and targeted but generally have higher installation and, if pressurized, energy costs.

Figure 5.3 Irrigation provides the life blood to many economies in arid zones as the contrast between the irrigated zone and barren hillsides near the River Nile at Luxor exemplifies.

Photo: Colin Chartres

Furthermore, some broad acre grain crops are not well-suited to drip irrigation because of cultivation requirements, and the large areas involved lead to prohibitive installation costs. Other critical factors to consider are the amount of reuse of excess water applications, and the amount of evapotranspiration (ET). ET includes the water transpired through the plant and evaporation from the soil surface. The key challenge for agriculture, whatever the method of water application/use, is to maximize the crop/animal production per liter of water evapotranspirated. Clearly the more water that is transpired through the crop, the higher the potential crop and water productivity. The more evaporated from the soil surface, the converse is likely to apply. If ET can be reduced and yet yields increased or maintained, water withdrawals for agriculture can also be reduced, leaving more water for other users and the environment. This entire area of

Figure 5.4 Gravity-fed irrigation systems are used extensively around the world. This example from Australia shows the feeder channel and laser leveled bays across which the water can be evenly distributed.

Photo: Colin Chartres

managing withdrawals, allocations, and developing water budgets for entire basins is a critical one for water management and is touched on again later in this book from the point of view of water governance.

At the global level, currently, about 7130 km³ (5.8 billion acre-feet) of water are evapotranspirated annually by agriculture.[3] This is equivalent to over 50 times the storage held behind the High Aswan Dam in Egypt, one of the world's largest dams. However, currently about two thirds of the world's agricultural production is based on rainfall rather than irrigation (see Figure 5.5). With population growth and changes in diets, it is estimated that food production will have to double by the year 2050. This increase in food production will need to come from both increasing irrigation area and productivity and from rainfed farming. Depending on the degree of soil fertility decline and achievable levels of water productivity, it has been estimated that up to almost twice as much ET will be needed to increase food production to required levels by 2050 (see Figure 5.6).

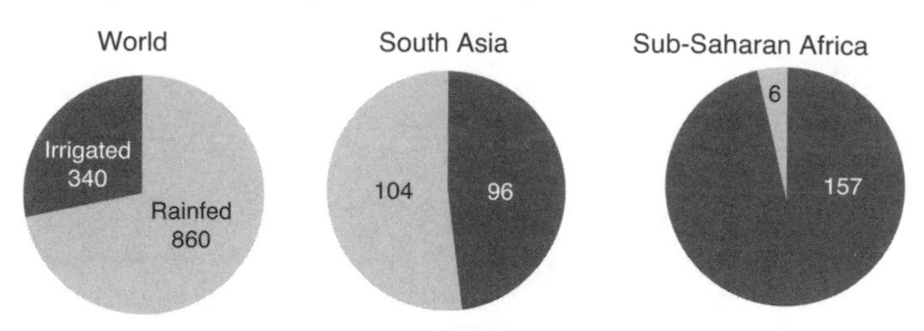

Figure 5.5 The proportion of harvested area that is irrigated

Source: Comprehensive Assessment of Water Management in Agriculture, 2007

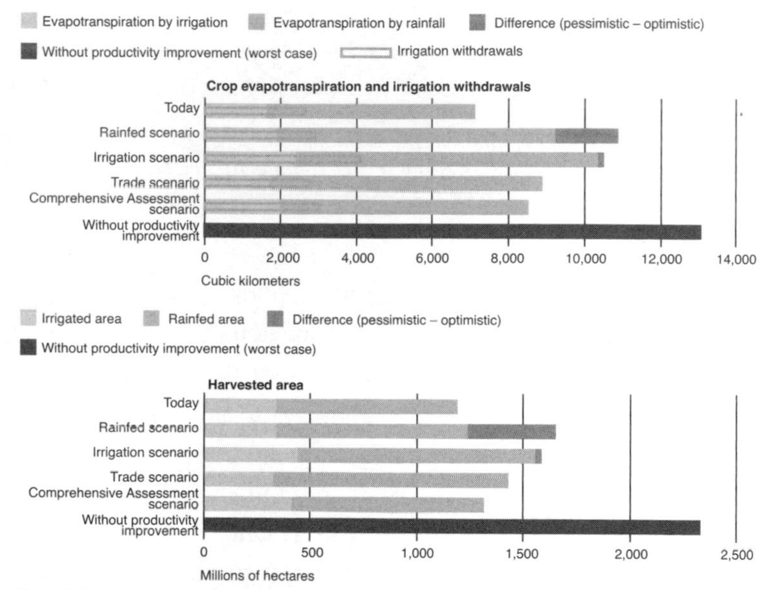

Figure 5.6 Water and land requirements to meet food demand in 2050 under different agricultural scenarios

Source: Comprehensive Assessment of Water Management in Agriculture, 2007

McKinsey and Company, a global management consulting firm, has indicated that over the next 20 years that water withdrawal requirements for agriculture will need to be increased by half as much as is used today if there are no efficiency gains (see Table 5.1).[4] McKinsey and Co. also emphasized that centers of agricultural and thus water demand are primarily in India, sub-Saharan Africa, and China. In India, they indicate that a gap of 50% between water supply and water demand will emerge by 2030.

Table 5.1 *McKinsey and Company—Projections of Increasing Water Demand by 2030**

	India	China
Projected water withdrawals in 2030	1,498 (1,215)	818 (663)
Projected 2030 supply	744 (603)	619 (502)
Water gap between demand and supply	50%	25%

*Numbers are in billions of cubic meters; equivalents in millions of acre-feet are in parentheses.

Whether or not the absolute increase in water requirement for agriculture in 2030 is as high as twice current consumption or less is immaterial given the magnitude of the task in providing such quantities. In contrast, water volumes used in industry are relatively small in comparison with agricultural water use. Nevertheless, industrial water withdrawals account for 16% of today's global demand and is anticipated to grow to a projected 22% in 2030. The growth will come primarily from China (where industrial water demand in 2030 is projected at 265 billion m^3, or 215 million acre-feet, driven mainly by power generation), which alone accounts for 40% of the additional industrial demand worldwide.

Demand for water for domestic use may decrease slightly as a percentage of total water use, from 14% today to 12% in 2030, although it will grow in specific basins, especially in emerging markets. Thus by 2030 to 2050, we may see a diminution in the proportion of water available to agriculture in the order of 5–10% of total available water withdrawals unless more supply can be made available. The latter is unlikely in most physically water-scarce countries. Additionally, the key issue then becomes how can we increase the productivity of water use in agriculture, with the critical challenge being to double production while withdrawing the same amount or less than we do at present? This challenge will be one of the most critical challenges faced by the world this century. Failure to meet it will mean malnutrition and potential starvation for many of the world's poor.

It is in Africa where the relationships between individual livelihoods and agriculture are most evident. In several of the smaller economies of the region, annual GDP is correlated with rainfall variability, thus demonstrating the significance of agricultural production to economic wellbeing (see Figure 5.7). A more worrying trend confronting Africa is that crop yields have shown little, if any, increase over the last 50 years. Thus while the "green revolution" transformed agricultural productivity in much of the rest of the developing world, for some reason, it passed by Africa, or at least poor African smallholders (see Figure 5.8). However, it has been argued that there is cause for optimism based around the Millennium Village Concept that is being developed in several African countries.[5] This concept involves a comprehensive strategy for rural development and short-lived subsidies that enable small holders to escape from the poverty trap.

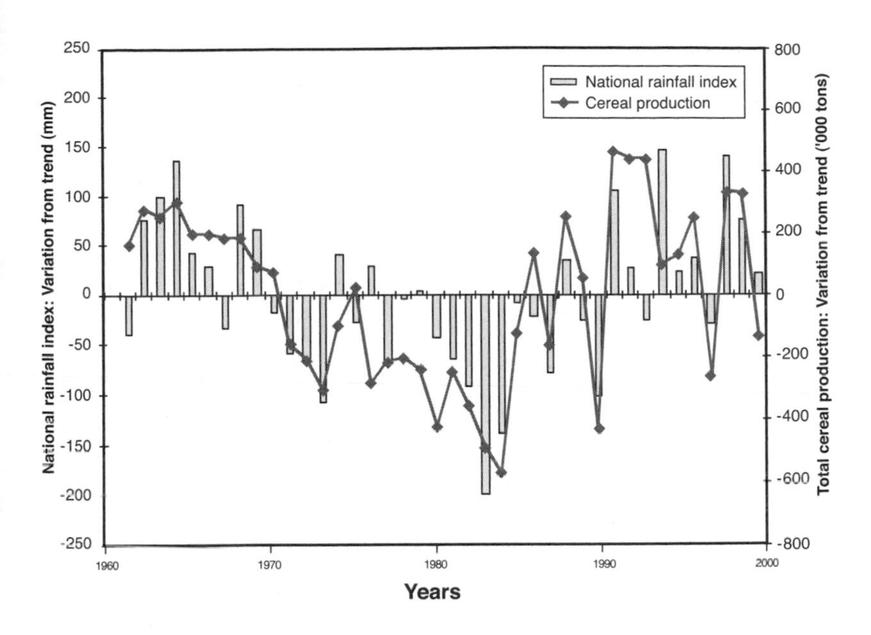

Figure 5.7 The relationship between rainfall and cereal production in Burkino Faso, West Africa (based on Food and Agriculture Organization of the United Nations statistics

Source: Comprehensive Assessment of Water Management in Agriculture, 2007

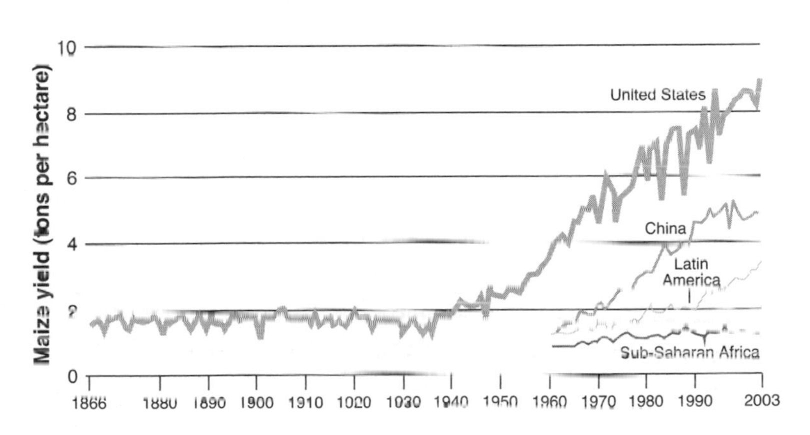

Figure 5.8 Yield growth rates from the mid-nineteenth century

Source: Comprehensive Assessment of Water Management in Agriculture, 2007

If we examine scenarios for agricultural water use in India, a business-as-usual approach to water management leads to some pretty grim forecasts from a range of scientists and water planners (Table 5.2). This table has a lot of data, but its careful analysis helps us understand how population growth and dietary change in India will drive up water demand. India currently has available about 690 km^3 (559 million acre-feet) of utilizable surface water resources and a further 396 km^3 (321 million acre-feet) of utilizable groundwater on an annual basis. This is unevenly distributed across the north and south of the country. In 2000, the total water demand was 680 km^3 (551 million acre-feet). Table 5.2 compares a business-as-usual scenario for 2025 and 2050.[6] With projections compiled by the Indian National Commission on Integrated Water Resources Development for 2050, the projections of water demand range from 822 km^3 (666 million acre-feet), not including industrial demand, to 1069 km^3 (887 million acre-feet). This latter figure is almost the same as predicted total utilizable supply. Furthermore, if the Indian economy has a high growth rate between now and 2050, the additional water requirement for the domestic and industrial sectors will exceed the estimated water reallocation from the agricultural sector. While the projections might suggest that India will cope with its water requirements until at least 2050, there is major concern that geographic disparities will require the National River Linking Project, mentioned in Chapter 2, "From Abundance to Scarcity in 25 Years," to shift large quantities of water from the rivers of the north to the drier south. Similarly, India will also have to invest heavily to ensure that currently underutilized supply (the "water gap" described by McKinsey and Company) is developed. This will, according to some commentators, lead to the loss of environmental flows and consequent biodiversity declines, the transfer of pollution from one system to another, and potential social unrest in those basins losing water—not to mention the enormous cost. As yet, not enough is understood about how climate change will impact both the flow of the large northern rivers nor rainfall in the south.

Table 5.2 *Summary of the Key Drivers and Water Demand Projections of Business as Usual (BaU) and Other Scenarios for India*[7]

Drivers	Unit	2000[i]	BaU scenario projections[ii]		NCTWRD high demand scenario[ii]		Seckler et al.[ii]	Rosegrant et al.[ii]
Population	Million	1007	1389	1583	1383	1581	1273	1352
- Urban Population	%	28	37	51	45	61	43	43
Total Calorie supply/person/day	Kcal	2495	2775	3000	-	-	2812	-
- % of food grains	%	65	57	48	-	-	58	-
- % from non-grain food crops	%	28	33	36	-	-	32	-
- % from animal products	%	8	12	16	-	-	11	-
Food grain demand/person/year	Kg	172	166	152	210	284	188	183
Total grain demand/person/year	Kg	200	210	238	231	312	215	215
Net sown area	Mha	142	142	142	144	145	-	-

Table 5.2 *Summary of the Key Drivers and Water Demand Projections of Business as Usual (BaU) and Other Scenarios for India*[7]

Drivers	Unit	2000[i]	BaU scenario projections[ii]		NCTWRD high demand scenario[ii]		Seckler et al.[ii]	Rosegrant et al.[ii]
			2025	2050	2025	2050	2025	2025
Net irrigated area	Mha	55	74	81	67[iii]	93[iii]	-	-
- from groundwater	Mha	34	43	50	34[iv]	42[iv]	-	-
Gross irrigated area	Mha	76	105	117	98	146	90	76
Irrigated area of grains	Mha	54	59	63	69	102	61	51
Rain-fed area of grains	Mha	69	62	57	70	57	61	69
Total grain availability/person/year	Kg	208	213	240	242	312	216	206
Net irrigation requirement	Km^3	245	313	346	359[iv]	536[iv]	323	332
Irrigation efficiency - surface water	%	30-45	35-50	42-60	50	60	-	-
Irrigation efficiency - ground water	%	55-65	70	75	72	75	-	-

Table 5.2 *Summary of the Key Drivers and Water Demand Projections of Business as Usual (BaU) and Other Scenarios for India*[7]

Drivers	Unit	2000[i]	BaU scenario projections[ii] 2025	2050	NCTWRD high demand scenario[ii] 2025	2050	Seckler et al.[ii] 2025	Rosegrant et al.[ii] 2025
Total irrigation demand	Km^3	605	675	637	611	807	702	741
- from groundwater	Km^3	272	304	325	245	344	-	-
Irrigation for grain crops	Km^3	417	398	351	428	565	-	-
Domestic water demand/person	m^3/day	33	45	64	45	70	31	31
Industrial water demand/person	m^3/day	42	66	102	48	51	55	
Total water demand	**Km^3**	**680**	**833**	**900**	**773**	**1069**	**811**	**822**

[i]Data for 2000 is from various publications of Government of India.

[ii]BaU, NCIWRD, Seckler et al. and Rosegrant et al. Information is compiled from GOI 1999, Amarasinghe et al., 2007, IWMI 2001, and Rosegrant et al., 2002, respectively.

[iii]Estimated with cropping intensities: 141% in 2025 and 155% in 2050.

[iv]Estimated with percent from groundwater irrigation 50% in 2025 and 43.7% in 2050.

It can, however, be argued that humans have an innate ability to innovate and adapt to changing circumstances and that by 2050, the likelihood of India running out of water will be minimal. Undoubtedly there are numerous ways in which agriculture can be more productive and produce more food from less water. However, this often requires not only the pessimistic predictions of doom and gloom unless something is done, but also a reform and incentive framework that facilitates innovation. There is considerable doubt in our minds whether most countries actually have such a framework planned or in place. So given the consensus among scientists and planners and the potential significant economic costs of large scale interbasin water transfers (as envisaged under the National River Linking Project) it is very clear that India has to act sooner rather than later to avert the consequences of doing nothing.

Recently, McKinsey and Company has considered the marginal costs of a wide range of interventions aimed at increasing water productivity in India and a range of other countries. What is encouraging is that many of the simple and feasible interventions, such as improving irrigation technologies, improving drainage, better fertilizer balance, and no till farming that can be delivered over large areas, are the lowest cost interventions.

Why Is Irrigation So Important?

Using Australia as an example, the following facts from the Murray-Darling Basin Commission illustrate just why irrigation is such an important component of agriculture and food production:

> "For Australia as a whole, the area of crops and pastures irrigated, 2,069,344 hectares (5,113,460 acres), is a minute proportion (0.4%) of the total area of land in agricultural holdings, 465,953,718 hectares (1,151,396,710 acres). It is only 11.7% of the total area of crops and pastures. However, the

> value of irrigated production is out of all proportion to the area of land involved. One source has estimated that irrigation accounts for between 25 and 30% of Australia's gross value of agricultural output. This puts the figure at up to approximately $7.2 billion. On top of this, there is an estimated four-fold multiplier beyond the farm gate."

At the global level, as demonstrated in the Comprehensive Assessment of Water Management in Agriculture, three quarters of the additional food demand in the coming 40 years can be met by improving the productivity of existing irrigated lands. However, at the regional level, opportunities to do this will be fewer, for example in China and Egypt, where yields are already quite high compared with South Asia where yields are much lower. So in reality, there needs to be an expansion of the area of irrigated land as well as productivity increases. Therein lies a paradox, in that in Asia a recent International Water Management Institute/Food and Agricultural Organization of the United Nations report[8] indicates that although Asia holds about 70% of the world's irrigated area, there is relatively little arable land left that could be developed for irrigation. In South Asia, 104 m ha (262 m acres) out of 200 m ha (494 m acres) are under irrigation (see Figure 5.5). Furthermore, in South Asia and parts of China, there has been very significant expansion of groundwater irrigation over the last 30 years, with many major groundwater systems consequently under unsustainable extraction regimes. Thus, with the exception of some smaller areas in South East Asia (Cambodia, for example), increasing irrigation area will not be the major pathway to food production increases in this region of the world. Increasing water productivity in existing irrigation systems, is, however, a different matter and has to be one of the solutions required.

Africa presents a different picture and has more scope to increase irrigation with only 6 m ha (15 m acres) out of 163 m ha (403 m acres) currently irrigated. However, international financing for irrigation development has reduced dramatically over the last 20 years based on a perception that the green revolution has dealt with food production issues once and for all and more recently, on a possibly justifiable perception that large surface irrigation schemes have not been very successful. (Hence, the significant increase in privately financed groundwater irrigation schemes observed in South Asia and China.)

Thus, it is clear that there is significant potential to increase the area of irrigation in Africa if we can find ways of changing perceptions about its feasibility. To do this will require more ancillary investment in the areas that underpin the viability of irrigation systems, including education and training, system management (see Figure 5.9), and improved access to markets and ways of cooperatively marketing produce grown by poor farmers with limited land (smallholders).

Figure 5.9 Wetlands offer scope for agriculture, particularly as annual floods recede. The challenge is to develop management practices that harmonize agriculture and biodiversity conservation.

Photo: Colin Chartres

Virtual Water and Water Footprints

In 2008, in the Stockholm City Hall (the same venue where most of the Nobel prizes are awarded) Professor Tony Allan from King's College, London, was awarded the 2008 Stockholm Water Prize. This prestigious prize was presented to him for his ground-breaking work on the concept of Virtual Water (also referred to as embedded, embodied, and hidden water). Virtual water can be defined as the water that is used to produce a food commodity. For example, it takes about 1000 liters (over 250 gallons) of ET to grow a kilogram (2.24 lbs) of wheat as a rule of thumb. The concept of trade in virtual water then refers to water-rich countries exporting agricultural commodities with a high water requirement to water-scarce countries. However, some economists are very skeptical of the concept of virtual water, preferring a framework that incorporates the value of water in any product and assumes that markets will then decide what trading should occur. To do this, however, also assumes that we have available tools that can cost water in terms of its delivery to the user and all externalities (such as, environmental damage, salinity, fisheries degradation, and so on) associated with its use. As yet, this has been very difficult to achieve.

If wheat with its virtual water content is then made into pasta, a further volume of water is then utilized in the factory process. When all the water used to develop a foodstuff is then summed up, we can see that the one kilogram of flour or pasta on the supermarket shelf may have had very thirsty origins. The concept of a water footprint is an attempt to sum up all the water that an individual, a particular product (such as a bottle of soda), a city, a mine, or a factory requires and the area required from which to harvest this water. This total amount of water is then described as a water footprint, in a similar fashion to the definition of a carbon footprint. The smaller the footprint, the more productive and environmentally sustainable the enterprise.

The virtual water concept leads to the question in water-scarce countries as to whether water-intensive crops should be grown there or whether such commodities should be imported instead. In principle, such strategies then allow for scarce water to be diverted to higher value uses and increase overall global water productivity. In 1995, without trade, de Fraiture estimated that irrigation withdrawals would have been 11% higher to produce the same amount of food. In practice, concerns about national food security, sovereignty, trade regulation, and sometimes kneejerk responses that turn off supply from exporting countries in times of potential food shortage mean that governments are loathe to rely on international trade for key commodities. However, in some cases they have no choice. In the Middle East and other regions, such as Korea, current concerns about water scarcity and food security drive national policies, and fear of too much reliance on open trade as a means of securing food has been the basis for the overseas land grabs referred to elsewhere in this book.

A few countries, notably, the United States, Argentina, France, and Australia are major virtual water exporters, and a significant number of others are classed as minor virtual water exporters. However, what is notable is that apart from the expected virtual water importers in the Middle East, virtually all the countries of sub-Saharan Africa (with the exception of Cote D'Ivoire) are virtual water importers. This figure exemplifies the fact that so many African countries are dependent on imported grains while in many cases they could actually be net exporters of foodstuffs. To achieve this status, however, they have to overcome economic water scarcity constraints, improve productivity, and continue to lobby for greater world market access.

This point is very significant in that it exemplifies some regions' reliance on "western" food producers and unless overcome will be a major impediment to economic growth and development in, at least, the water-rich countries in sub-Saharan Africa. One of the scenarios examined by the Comprehensive Assessment of Water Management in Agriculture was the extent to which enhanced trade in foodstuffs could help feed the world's growing population. This is explored in the following section.

How Can We Achieve More "Drop Per Drop"?

While food scarcity may see the financial pendulum swinging back to investment in surface irrigation, increasing the productivity of water in irrigated and rainfed agriculture must be considered as a sound strategy by governments. There are a number of options over the continuum from rainfed to irrigated lands identified in the Comprehensive Assessment on Water Management in Agriculture, including

- Supplemental and deficit irrigation
- Maintenance of soil fertility
- Small-scale affordable management practices for storage
- Delivery and application
- Zero or minimum tillage
- Breeding and biotechnology for fast ground cover
- Reduced susceptibility to drought

The key questions are how can more crop per drop be achieved in the face of a number of limiting factors including low soil fertility, risk from highly variable rainfall, poor water management practices, and often limited human capacity manifest in poor farming communities? Clearly a number of actions beyond the provision of technologies are important.

If these measures are combined with increasing the productivity of irrigated areas, development of new irrigation schemes, particularly in Africa (ensuring that agricultural trade within and between countries is not impeded—reducing food demand by influencing diets and minimizing post-harvest losses), they can have a major impact on improving food production.

The Comprehensive Assessment of Water Management in Agriculture modeled a global scenario that assumed zero growth in irrigated areas and showed that rainfed agriculture could meet the increase in food demands projected for 2050. This optimistic scenario had cereal yield increase by 72% and harvested rainfed area increase by only 7%. This scenario would lead to sub-Saharan Africa, Latin America, and most of Asia being largely self-sufficient in producing major food crops, with the exception of maize in East Asia. A more pessimistic scenario that included only modest yield improvements projects the rainfed area increasing by 53% to meet future food demands. However, this scenario required countries without available land and/or reliable rainfall to significantly increase food imports. Under this scenario world trade in food would expand from 14% of production today to 22% in 2050. Potentially, there is the land available to increase production even under the pessimistic scenario, but the extent to which it has better than marginal production capability and the extent to which climatic factors including rainfall uncertainty would limit potential production is uncertain. Furthermore, given climate change scenarios and information regarding underutilized lands, the possibility of expanding and increasing rainfed production may be much less in Africa than in Eastern Europe.

The next question to ask is whether the modeled changes could realistically be achieved given current climate and other factors? Both optimistic and pessimistic scenarios call for substantial increases in soil water consumption. For the optimistic scenario with little expansion of rainfed area, improved water management, including small amounts of supplementary irrigation, is a prerequisite for yield increases. This will require more ET, part of which can be offset by increasing water productivity by improving the harvest index (ratio of yield biomass to the total cumulative biomass at harvest), by reducing losses from soil evaporation, or by increasing transpiration while reducing evaporation.

The optimistic scenario sees rainfed cereal production improve by 72% with a concomitant increase in water productivity of 32% compared with 20% and 10%, respectively in the pessimistic scenario. In total, compared with the year 2000, each scenario would use an additional volume of water of 2150 km^3 (1.7 billion acre-feet) and 3850 km^3 (3.1 billion acre-feet), respectively. It is highly probable that increase of water consumption of these magnitudes will have major impact on river flows and groundwater recharge in most areas, so the optimum pathway would be to follow the optimistic scenario.

Obtaining a 32% increase in water productivity in rainfed agriculture is, however a daunting task. It will require major rethinking of the way in which governments and society view agriculture in general. It will require much better knowledge about soil types and soil water and soil structural management to maximize water availability. It will also require major investment in capacity building of farmers and other players in the food production chain as well as require governments to pay very significant attention to developing policy settings that promote agricultural productivity increases that build on knowledge and capacity development as opposed to subsidizing particular aspects of agriculture such as fertilizers. It will also be imperative that investment levels for research and development are stepped back up to levels that they were at the time of the green revolution in order to overcome the very significant challenges facing yield and productivity improvements across the agronomic-water interface. Finally, it will require innovative thinking by all concerned to optimize the best productions systems for given environments that also pay attention to maintaining the provision of ecosystem services.

Urban Water Reuse for Agriculture

Approximately 80% of water entering cities exits as wastewater. This water is, of course, polluted with a range of domestic and industrial contaminants. However, the majority of contamination is from sewage, which contains large amounts of potentially valuable nutrients and organic material. In fact, in some areas farmers value partially treated or untreated urban effluent as a free source of fertilizer and a source that is usually available 365 days per year. However, unless treated to some degree, such waste waters are also a major hazard to agriculture, the environment, and public health. Poor farmers, who see such water as a free access to irrigation and nutrients put their own health at considerable risk just by handling the water because of its high fecal coliform bacteria, viral, and parasitic worm content. Consumers who eat vegetables grown in such water are similarly at risk if the vegetables are not adequately washed or if the water used for their production contains heavy metals or persistent organic pollutants that can be absorbed by the plants. Receiving waters are at risk because of both the high nutrient loads that literally poison aquatic life and cause eutrophication (the process in which fresh waters become increasingly rich in nutrients such as nitrogen and phosphorus) and consequent algal blooms that feed on these nutrients and reduce water oxygen levels to the extent that fish may die.

In some cases, significant quantities of pharmaceutical products are found in sewage. These contain endocrine disrupting chemicals that may cause mutation and other impacts on the fauna and flora in lakes and rivers or even nearby seas. While sewage treatment facilities in the developed world have been able to deal reasonably successfully with the majority of these impacts, this is far from the case elsewhere. Consequently, a major challenge for water managers in the current century is to develop new sewage treatment systems that reduce significant health risks from discharge of wastes into the environment and allow agriculture to capitalize on the potential nutrient and irrigation water benefits that undoubtedly exist in urban effluent. Unfortunately, the continuing separation of water supply and sanitation sectors from water resources management sectors in most countries means that in the developing world this is a challenge that has not been adequately tackled as yet. Further discussion of wastewater reuse is in Chapter 7, "Integrating Water Planning and Management."

Conclusion

It seems inevitable that in spite of its importance to food production, agriculture will see increasing competition from other users for precious water supplies. While some governments may want to legislate and regulate to protect agriculture's share of water, the Comprehensive Assessment and the McKinsey Charting Our Water Future study both suggest that there are a number of key steps that can be taken to increase the productivity of agricultural water. While a number of these are physical in nature, many relate to policy changes, which are explored in more detail in later chapters.

This chapter examined critical issues with respect to water use in agriculture and how agriculture can facilitate the major production increases required to feed the world's growing population—and in particular the world's poor. It seems clear that in Asia there is limited scope for increasing the area of irrigated agriculture, although this may be a good way forward in Africa. The data presented suggest that rainfed agriculture, because of its widespread predominance, will have to significantly increase its productivity. However, the increases are potentially attainable in terms of available land and water resources. The sheer numbers of people to be fed by 2050 and the competition for water from other sectors of the economy mean that the challenge for the agricultural community will be immense, and this needs to be recognized by governments across the world in terms of their preparedness to invest in agriculture and natural resource management at hitherto unforeseen levels. The ultimate challenge for agricultural water use will be whether productivity increases can be achieved that enable adequate food and nourishment for a population of 9 billion in 2050 if agricultural water supplies are reduced on average by up to 10% by competing uses and climate change. This is a somewhat scary scenario but one that is highly likely. One of the key research challenges of the current decade is to determine more precisely those areas at greatest risk of reduced agricultural water allocations so that the potential solutions described earlier can be brought into effect.

chapter six

Water, Food, and Poverty

The relationship between water, food, and poverty is complex. However there are two ways in which the connections can be viewed. If we accept that to be free of poverty is to be free of hunger and experience a degree of well-being that enables a person to work, "food security"—ensuring enough food for all—is a precondition to poverty alleviation. A food-secure world requires water to be used more efficiently to end hunger and malnourishment. We also know that for a large number of the poorest people who live by farming, moving from farming-for subsistence to farming-for-profit is not easy. Access and availability of water, one of the basic inputs required for agriculture, can make a huge difference to the productivity of smallholder farms and help them escape mere subsistence agriculture, making farming profitable.

There are more than one billion people, predominantly in Asia, who live on incomes below the poverty line, at risk of further malnutrition. In Africa, the inability to grow enough food or provide access to it results in millions facing starvation every year. There is considerable overlap between the group of people who need to grow enough for survival and those whom depend on farming for livelihoods and also suffer from chronic poverty. In the next 40 years, we will confront the challenge of having to feed another 2.5 million people with less water for agriculture than we have now. The rising cost of food has increased the number of people who suffer from chronic hunger. This chapter reviews some of the most important issues that link water, food, and poverty and are at the focus of efforts to improve food security as well as the productivity of agriculture. One of the fundamental questions that links these three elements has to do with our relationship with the environment and the ways in which we use natural resources to support our existence as well as the dangers of overexploitation, both for nature and humankind.

In this discussion of water and food, there are several global issues that contribute to poverty, particularly the changes being brought on by climate variability, decreasing investment in agriculture and water development, and population pressure. All of these relate almost directly to the challenge the global community faces, in providing enough food for all. In this chapter, we look at the big picture of growing enough food to prevent hunger, and then look into the specific problems faced by farmers in different contexts. The productivity and profitability of farmers relying on rainfed, canal, or groundwater irrigated agriculture differ greatly and, therefore, the conditions that contribute to poverty in each vary, with either the environment or institutions playing a larger role. For instance, the development of irrigation infrastructure to meet the food needs of growing populations in Asia has provided several lessons on the role that irrigation has played in poverty reduction, yet there still remain large pockets of inequity within these systems.

On the other hand, the groundwater boom has shown new ways in which the distribution of and allocation mechanisms for water has changed because of its unregulated use in agriculture, and has had a positive impact on poverty but also raised concern that it may not be sustainable in the long term unless practices change. An important issue that is consistently brought up in international development is that of gender and the problems that women in agriculture and water management face as farmers. The debate revolves around the problem of access, participation, and rights—all of which, when not addressed, affect the overall productivity of agriculture and hence economic development. Long-term impact on poverty through technology and development of water resources can only be successful if there is a concentrated effort to target programs and investment into poor areas and for smallholder farming. This should not be seen as a purely welfare-oriented approach, but also as a means for overall economic development.

The Big Picture—Farming and Poverty

The vast majority of poor people living in the developing world depend on agriculture to make a living. In terms of actual numbers, about 70% of the poorest live in rural areas where there is little other than agriculture as a means of employment. Farming is known to be one of the most precarious ways to earn a living, however. The difficulty of being a farmer is well-known—apart from having to contend with nature, the cost of basic inputs such as fertilizer, seeds, and water vary greatly depending on where in the world you live. In Asia, for instance, access to water even in areas where physical scarcity is not a problem is often still difficult due to inefficient and inequitable institutions stemming from organizations and policies that determine how water is to be shared. Power and politics often affect access, especially for vulnerable groups such as women, children, the aged, and marginalized indigenous and other communities.

In Africa, on the other hand, the lack of sufficient infrastructure and the poor design of many water delivery systems have meant that vast productive areas of the continent that could have access to water have been left out. Investment in water development, the construction of dams, canals, and reservoirs or even small scale storage structures such as ponds has been less than adequate for the needs within the continent, especially the countries struggling to be food-secure. Water storage per person in Ethiopia is about 14 m^3 with many other countries only marginally better. Compare this with the thousands of cubic meters per person in Australia and the United States! On the brighter side, however, reports by the World Bank and FAO have said that an additional 400 million hectares of land spanning 25 countries in Africa are suitable for farming. It is undisputed that a reliable source of water for a poor farmer can make a whole lot of difference. Making water available makes it possible for poor farmers to do more, grow more, and therefore be more likely to escape the cycle of farming-for-survival. For poverty alleviation, this is crucial.

In spite of the availability of resources in Africa, overall it seems that the task of reducing poverty is only getting harder. The global community is now facing challenges that affect us all, the rising cost of food and fuel, the pressure of inadequate resources to grow more food across the world that is leading to the "food land grab," and the looming presence of climate change. All of these factors affect the poor more acutely. Farming is becoming harder as demand for scarce water rises and as climate change affects the availability of water, making rain less predictable or drought and flooding more common. In parts of the world such as Bangladesh, which is predicted to experience worse flooding, and many low-lying island states of the Pacific and Indian Oceans where sea level rise is a major threat to their existence, poverty is likely to increase. Countries who are resource-limited and yet financially rich have recently begun to try to buy large tracts of land in resource-rich but financially poor countries in order to grow food. This has resulted in some developed countries have been recently accused of engaging in "agri-colonialism," which is being thought of as another form of imperialism, as the poor in those countries are denied ownership of their own land for the sake of the profit that can be made by these investments.

Nature and Poverty

What we take from the environment and use to support life (things like clean water and nutrients), the processes that control climate, and the values we attach to nature are collectively referred to as *ecosystem services*. These support the activities that are crucial for survival, most important of which is the production of food. The relationship between vulnerable environments and poverty is a close and strong one. First, as populations grow and the demands placed on the environments to grow food and support human activities rise, the resource base eventually starts to degrade. It has been observed that environments that support large populations are often ecologically vulnerable. Second, the degradation of ecosystem services can result in poverty as the over-exploitation of resources can reduce the potential of the land to grow food, leading to eventual impoverishment of the people who live off the land.[1] Examples of the two-way process can be seen in parts of the world where there is a great dependence on the environment and overuse of resources and where the poor are caught in this dangerous cycle. In Kenya's Mau forests the ecological disaster caused by clearing and felling the forest is said to be responsible for some of the worst drought that country has faced. The rivers that come from the forests have been greatly affected by the overpopulation and overuse of the forest land, affecting the poor farmers downstream: Approximately 10 million people depend on the rivers that originate from the forest. Jared Diamond's excellent book *Collapse*[2] provides similar compelling examples of how several relatively advanced societies collapsed due to overexploitation of their natural resource bases.

Estimates suggest that about 60% of Asia's poorest live in environmentally vulnerable areas, and about 50% do in sub-Saharan Africa.[3] These conditions are most often in mountainous regions, uplands, flood-plains, and forests. Food production in these regions where agriculture is mostly conducted by smallholders and dependent on rain for irrigation is often done at a small scale and for subsistence. Poor and vulnerable farmers in these regions are often left out of general development and investment aid, often due to their location or the significantly greater efforts required to improve the productivity of these regions. As a result, the efficiency and productivity of these regions is low, and in most countries it is evident that the poverty levels of the rainfed or "unirrigated" areas is far greater than in the plains where large formal irrigation schemes exist. Large and persistent gaps in terms of poverty levels have also caused large numbers of people to migrate out of these regions.

From a poverty perspective, the main question it leaves us asking is how can irrigation improve the productivity of smallholder farming? The importance of smallholder farming to food production is also something that often gets overlooked. About 85% of the world's farms (445 million of 525 million farms) are less than five acres and contribute to a large percentage of the cultivated land in many countries. Because the vulnerability of the environment and its impact on the poor is the greatest challenge in these regions, the main purpose of providing supplemental irrigation, sources of water to supplement rain, is to help insure against the shocks and stresses to agriculture that come about from other environmental factors. Some of the most interesting and productive developments in improving agricultural productivity are experiments involving small-scale technologies that have served to supplement rainwater for farming. At a glance, these may not seem like the most innovative or groundbreaking technologies or ideas. However, simple methods to store water in ponds or tanks have been sufficient to supplement water sources for smallholder farmers who live in areas where they are dependent on increasingly unreliable seasonal rain for farming (see Figure 6.1).

Figure 6.1 In Ethiopia, traditional irrigation schemes have shallow, hand-dug channels that convey the water to the fields. They demonstrate how low technology water storage and irrigation systems can have major health and livelihood benefits for communities, and thus provide "insurance" against climate risks.

Photo: Colin Chartres

Irrigation and Poverty—Learning from Asia

The importance of irrigation infrastructure to poverty alleviation has been acknowledged in Asia for a very long time—investments in large-scale irrigation projects in South Asia can be traced back to British colonialism. In the post-colonial period, investments in infrastructure including irrigation were influenced by modernization and dominant theories of development, and large infrastructure was seen to be symbolic of growth and development. Because many countries in Asia had, and still have, large agrarian sectors, irrigation is seen as a means not just to improve growth, but also to help improve the livelihoods of the vast populations of the rural poor who are involved in agriculture. In fact, irrigation has played a huge role in improving agricultural productivity in South Asia, contributing to improving the food security of the region: Between 1980 and 1995, India's net irrigated area increased by 30%; in Bangladesh irrigated area increased by over 50%. The growth in food production that marked this period is strongly attributed to the expansion of irrigation and the growth of irrigated agriculture, which was accompanied by the use of higher yielding varieties of crops and fertilizer.

In this period, policies and interventions to improve agricultural production were also accompanied by complementary programs to improve rural development and resulted in a big push to develop roads, health centers, educational and training facilities to improve extension services, electricity, telephones, and services like credit. Hence during the green revolution, investments in irrigation were complemented by a general improvement in rural infrastructure. This was a particular feature of the period and made it difficult to distinguish how irrigation on its own impacted poverty, blurring the natural tradeoffs of investing in one type of infrastructure over another. Following this period and the initial expansion of irrigation in Asia, the high environmental costs of unregulated water use lead to increased salinity, water logging and, eventually to stagnant or diminishing yields in some areas and a loss of livelihood for many poor people. The environmental costs of irrigation constitute some of the most critical issues to irrigation development. There are two positions to the debate on the role of irrigation on poverty. Proponents of irrigation's role in poverty reduction often site the importance of reliable access to water in poor rural areas, which comprise much of the developing world. Reliable access to water, the key resource required for agricultural production, can help poor farmers increase their production of food, intensify cultivation, diversify crops, and experiment with new technologies that will improve their production and thus improve their standard of living. Opponents, on the other hand point out that the problems of environmental degradation that result from overextension of natural resources eventually have an impact on rural livelihoods that depend on agriculture. Again, the long-term effects of environmental degradation would then have the greatest negative impact on the poor, who are most heavily reliant on natural resources for their incomes.

Irrigation investment trends have reflected a shift in thinking on irrigation development, from the high-investment phase following the green revolution to declined spending on irrigation since the late 1980s. In the 1960s, the average rate at which global irrigated area rose in was 2%, in the 1970s it rose to 2.4%, but by the 1980s had fallen to 0.9%, mirroring the decline in public spending on irrigation infrastructure in developing countries over the same period. The causes of declined investment have been attributed to the high cost of construction and, related to it, the poor rate of recovery; environmental impacts; falling agricultural prices that resulted from the initial expansion, and the poor performance of irrigation systems. The poor performance of these systems in Asia has given rise to much debate on what needs to be done to improve the productivity of agriculture.

Research in canal irrigation systems have shown that the yields from irrigated areas are around twice as much as those from rainfed areas—not surprising as this indicates that farmers in irrigated areas can grow a more diverse range of crops and grow for more seasons. For the poor, landless population in these areas, there is a greater opportunity, and wages tend to be higher because the promise of irrigation means that there are also greater opportunities for employment on farms.

Although in many countries agricultural water is provided through large-scale public irrigation systems, there are many reasons why the performance of these has declined over the years. Most of the factors that have affected this decline have to do with the management of these systems and the ways in which they have not served the poor. The location of people along the canal system has a lot to do with how much water they may receive as well as its quality. In the typical discussion of equity issues in water management, farmers located at the tail end of canal systems, downstream, are often the recipients of irregular flows of water and also have to contend with the runoff from the land upstream. This means that the pesticides leached from the land in addition to sources of pollution such as excess fertilizer, sanitation waste from settlements and farms, as well as the waste from livestock altogether could affect the quality of production on land downstream. Another issue that often arises is the social dynamics at the local level. Power and politics also determine whether farmers are connected to the system. Large pockets of poverty persist in many large scale irrigation schemes, and this problem is often attributed to local politics or the marginalization of communities along cultural lines. The institutional arrangements that determine how water is allocated or shared are not exempt from power struggles, and not surprisingly, the size of land ownership will determine the wealth of an individual farmer and more likely the power that can be wielded in decision-making.

Groundwater and Poverty

In South Asia, the proliferation of groundwater irrigation, made possible by the availability of cheap, easily accessible diesel and electric pumps, has made it one of the fastest growing forms of irrigation. Groundwater irrigation in the region has often been described as being anarchic and "atomized," springing up through the work of individual farmers in contrast to centrally planned surface irrigation systems. The most interesting lesson from the rapid rise of groundwater irrigation is that it has put control of water into the hands of farmers. Farmers have then either used this water to grow more food or diversify their crops and also, interestingly enough, used it to sell water.

Water markets, that is the buying and selling of water resources, is a known phenomenon in many developing countries where access is otherwise limited. Those who can afford to purchase a pump that will help lift water from a canal or a well have the opportunity to control the resource that others do not. The costs of groundwater extraction vary, depending on the mode of pumping. In parts of India where cheap electric pumps are used and flat tariffs are charged for electricity, some farmers have found it more profitable to sell the water that they pump to other farmers without direct access to water than to actually farm on their fields.[4] And so not only has groundwater helped farmers gain access to water whenever they want (without having to wait for the prescribed times when water is released from through the canal systems), but also it has enabled some to develop a market around the buying and selling of water.

Groundwater use in South Asia has also been described as being democratic because of the way in which it has sprung up, giving farmers more choice.[5] The classic equity issues that crop up in canal and public systems such as not enough water for farmers downstream or the issues of corruption that may negatively impact some water users do not arise when farmers have total control and their own means for pumping. Another interesting poverty aspect of the groundwater boom is that groundwater markets may actually be serving to provide water to a greater number of farmers than canal systems are able to reach. For instance, poor farmers with limited access to canal water and who can grow for less time may have the option to buy water from groundwater sellers and grow for an additional season or they can use the groundwater as a risk mitigator to insure them when rains are erratic.

The problem of groundwater use in agriculture is that overexploitation of the resource has resulted in water tables reducing to very low levels in regions. Although groundwater is a hidden resource that many scientists say could be the answer to the world's water problems, unregulated use in areas where rainfall is low and where farming practices are not being planned for sustainable use could cause serious environmental problems, such as desiccation of rivers and drying of wetlands.

Women, Water, and Food

The question of ownership and control of resources matters greatly in water management and its effectiveness. When looking at the contributions of women to agriculture, it is hard not to notice the relative imbalance in their labor contributions versus their actual ownership and control over resources such as land and water.[6] Across Asia, women are involved in many aspects of agricultural production, such as planting or weeding, and also in the post-harvest sorting, drying, and cleaning of food crops. Yet women own or possess the rights to very little of what they help to cultivate.

> "In terms of the environment, women around the world play distinct roles in managing plants and animals in forests, drylands, wetlands, and agriculture; in collecting water, fuel, and fodder for domestic use and income generation; and in overseeing land and water resources. By so doing, they contribute time, energy, skills, and personal visions to family and community development. Women's extensive experience makes them an invaluable source of knowledge and expertise on environmental management and appropriate actions."[7]

Correcting these inequalities has been the inspiration for land movements and rights struggles across the world. Water management institutions such as water user associations and farmer organizations have often kept women out of key decision-making spaces.

The productive contributions of women's time and labor in food production also often goes unnoticed. For the majority of the poor living in rural areas, migration to the cities is fast becoming the first option to look for jobs and new sources of employment. In large parts of sub-Saharan Africa, the migration of men has meant that there is a disproportionate number of family farms being headed by the women who have been left behind. Women in rural areas produce between 60% and 80% of the food in most developing countries. With the migration of men, the central role of women farmers in food security has still to be recognized. The undervaluing of these contributions is evident in much of development policy that has not adequately targeted or included women. The most important of these is technology. The marketing of agricultural technology has always been gender-biased.[8] Assumptions about gender and the roles of women in agriculture has meant that little has been done to get women to use technology or to target technology toward them. Culture plays a role here, as in many places it is assumed that men are the ones who are somehow better skilled or equipped to use technology. However technology transfer is still seen as one of the most important aspects of addressing the poverty issue, particularly because of the large number of women who are poor and have been left out.

Despite the importance of technology and access to water, these elements together are still not enough to support women farmers. The issue of water management is also an institutional one, and as the shift to using water user associations as the place where decisions regarding water are made in many countries, it became clear that women were not able to participate in these spaces. Not only were they absent, but very often their voices were not taken seriously. The absence of women in these institutions, in the long run, threatens their effectiveness especially in places where there are a large number of women farmers. The participation of women in water user associations and other kinds of organizations where decisions on sharing water or other resources, globally, is poor.[9] While there are cases where women's voices have been heard and made an impact, these are not consistent and much more needs to be done to improve the culture of these institutions.

Water as Welfare

In most developing countries, governments play a role in the distribution of water. Poverty alleviating schemes such as food-for-work programs in India and other similar programs that are targeted to poor farmers unable to ensure paid work between growing seasons have often been based around water development. The focus on building water infrastructure, such as dams, canals, and tanks has been an interest for governments of developing countries for a long time. Programs that get farmers to contribute labor for the building and or maintenance of these structures have long been used as a way to provide wage labor for farmers while also contributing to the construction of public water schemes.

The importance of meeting the basic requirements for nutrition and well-being for those who suffer from hunger is the fundamental objective of those working to improve food security. While this is fundamentally the first objective of our efforts, long-term efforts at poverty reduction should be aimed at improving the access to and productivity of water in order for people to grow more and improve their incomes from agriculture. Welfare approaches, such as those just described have been hugely successful as a way to provide a safety net for farmers during times of stress or shock, however investment in agriculture needs to take a more direct approach to addressing poverty. The difference between the levels of assets among the poor who depend on farming for livelihoods indicates that there needs to be a more proactive effort to make resource allocation more equitable. This is not important just from the perspective of poverty and meeting basic food security; it is also important because of the extent to which productivity and the contribution of smallholder farmers is to food production worldwide and particularly in the poorest countries.

Conclusion

Farming across the world is a precarious occupation and one that many point out will become increasingly difficult with climate change. For the poorest who rely on farming for subsistence, and on whom many of us rely on to produce our food, water is a key element to survival. Efforts to improve the way in which water is used so that more food can be grown to feed the world's population range from technological solutions to move and lift water to physically scarce areas, policies that enable the virtual trading of water, or attempts to secure productive land to grow food in light of the climate's effect on environments. All of these relate to the challenges that face most countries in their attempts to reduce hunger. Global investment in agricultural water management and development has shrunk over the past years, despite the fact that investments in water development, for small-scale use such as simple technologies as well as traditional irrigation, have been shown to lift farmers out of poverty. Also, there is no guarantee that policies will affect those that need support the most, and entrenched politics and social differences play a very important role in access and rights to water resources. Although there are many examples of how, essentially, "privatized water supply" such as groundwater has made a huge impact on poverty with poor farmers in those regions being instrumental in developing the market in the absence of public provision, there are concerns over the unregulated rise of this phenomenon. As humans, our survival is dependent on the environment, and exploitation of resources will ultimately affect us all, with the more immediate and severe consequences being borne by poor communities.

The role of women in water management has long been overlooked partly because of negligence with regards to the number of women farmers, but also the difficulty of understanding how they can be supported. Because of the difficulty of accounting for their contributions to agricultural productivity and also the restrictions on securing rights to access to resources, we are terribly behind in our support to women farmers. The knock-on effect of these can be seen in many areas, where the status of women has been affected by a prolonged neglect of rights and access. The solutions, once again, cannot be merely technical and have to ensure that women producers of food also can participate in water user associations and spaces where decisions are made.

An increasingly complex question that arises in discussions of water management is the role of the government. As the provision of water becomes privatized or determined by decentralized bodies, it is becoming harder to ensure that the poorest people don't get left out. The question remains as to whether the government should play a bigger role in ensuring that water allocation is more equitable so that the poorest and most vulnerable groups are not consistently denied access. Governments have also used water management to help the poor by providing employment in schemes to tide farmers over during times of drought or crop failure. These programs have been invaluable as a means of security for farmers.

chapter seven

Integrating Water Planning and Management

"Water and air, the two essential fluids on which all life depends, have become global garbage cans."

Jacques Cousteau (1910–1997)

When we go to the toilet, a simple flush carries the waste away and out of sight. Few of us give it further thought. However, wastewater treatment is big business in the developed world and of increasing concern in the developing world, where investment in sewage treatment plants has been very limited. While most rivers in the developed world are now relatively clean and protected to a greater or lesser extent by water quality laws such as the EU Water Framework Directive,[1] the same is far from the case in many developing countries. Due to a combination of neglect, lack of investment, poor understanding, and lack of laws and regulation, many rivers in these counties are dangerously polluted. Furthermore, while untreated waste in these rivers in the past was to some extent dealt with by "the dilution is the solution to pollution principle," increasing water abstraction now means that many rivers are severely polluted. However, many such rivers including the Ganges in India and Yellow River in China are still the source of drinking water and agricultural water for millions of people. Inadequate water treatment means that millions are still at risk from both biological contaminants and industrial pollution including toxic heavy metals. Some rivers such as the

Buriganga[2] in Dhaka, Bangladesh, are little more than open sewers and have lost all their beneficial environmental services functions.

A key message of this book is that if we are to overcome the emerging world water crisis, water governance and management have to be viewed holistically across all sectors of the economy and environment. This is only just becoming the norm in some countries and is far from the case in many others—and in most places, putting it into practice is the biggest challenge. While drinking water supply and sanitation issues were for the majority of the last two centuries treated separately from agricultural water use, in reality, all uses of water are dealing with a single resource. Thus, it is critical that we start to view, govern, and manage water in an integrated manner following the principles of integrated water resources management (IWRM).

The Global Water Partnership[3] commented a few years ago that

> "IWRM is a challenge to conventional practices, attitudes and professional certainties. It confronts entrenched sectoral [pertaining to sectors of the economy] interests and requires that the water resource is managed holistically for the benefits of all. No one pretends that meeting the IWRM challenge will be easy but it is vital that a start is made now to avert the burgeoning crisis."[4]

Integrated Water Resources Management practitioners have been saying this for many years, but it is only relatively recently that governments and international financing and development agencies have started to contemplate how to do it. In the water supply sector this has been triggered partially by the realization that by careful catchment management, significant amounts of money can be saved that otherwise would have to be invested in very expensive water filtration plants. One of the aims of this chapter is to examine the interface between land uses, water quality, and water supply and sanitation in the context of IWRM.

A second aim of the chapter is to look at the critical issues of the drinking water supply in the developing world and consider how concepts of IWRM may be used to everyone's advantage by looking at how multiple use water systems can be developed to supply households with both drinking water and irrigation water and how agriculture can have a place in cleaning up polluted water.

Integrated Water Resources Management

Following the international conferences on water and environmental issues in Dublin and Rio de Janeiro held in 1992, the concept of IWRM has attracted increased attention. Unfortunately, however, IWRM as a concept has neither been unambiguously defined, nor has the question of its practical implementation been adequately addressed. Water has many uses and users. Users include the following sectors of the economy: agriculture, water supply and wastewater, mining, industry, environment, fisheries, tourism, energy, and transport. In broad terms, the principle of IWRM recognizes that all these users are essentially interdependent. Similarly, the practice of IWRM requires that all water management institutions and users of water are cognizant of the needs of each other and that governments provide a planning framework that enables and empowers water managers to make sensible decisions. Herein lies a catch. While the intent is good, many governments have either failed to provide a legislative and regulatory framework that supports the concepts of IWRM or have labyrinthine responsibilities for water scattered through numerous departments, which impedes progress. A classic recent example from Australia is that health department regulations have actually impeded water saving measures such as rainwater harvesting in urban environments. Elsewhere lobbying from vested interests has prevented legislators from effectively minimizing point source pollution from factories and intensive animal production facilities. Examples of this are common in the United States. Recently in the light of the 2009 outbreak of swine flu, CSE, an Indian non-governmental organization (NGO), pointed out that a company at the center of claims (which were probably false) regarding the origin of the flu was able to claim that its contracted farmers and not the company were solely responsible for the waste produced from intensive pig farms. This allowed the company to bypass environmental laws that mandated a company had to get rid of its waste, but only if the waste belonged to it.

Clearly, it is in everybody's interest that water catchments are managed to optimize both quantity and quality of water. In the context of IWRM, this means that water needs to be shared equitably between users and pollutants managed to keep contamination to a minimum. In most countries agriculture is not excluded from water supply catchment areas (watersheds), although wherever possible water supply is harvested from pristine and often forested catchments. As water catchment areas become increasingly developed and agriculture and other industries intensify, it becomes increasingly difficult to maintain water quality.

Drinking Water and Sanitation in the Developing World

In developed countries, having a piped water supply 24/7, of drinking quality, is taken for granted. However, the situation in the developing world at the turn of the century was somewhat the reverse with vast numbers of people without access to these basic facilities. Women were and are most affected by this, not only having to spend hours carrying water, but also being at risk of attack and rape when forced to venture away from their homes to answer calls of nature. Furthermore, many of the diseases responsible for infant mortality and adult debilitation are associated with inadequate water for washing and sanitation and poor wastewater disposal. For the very poor in developing countries, incomes just do not extend to affording even the most rudimentary water and sanitation connections. As indicated in the previous chapter, there are about 1.4 billion people now living on less than the new threshold of $1.25 per day. Until 2002, there was an overall decline in the number of rural poor (defined in this case by those living on less than $1 per day) from 28% in 1993 to 22% in 2002,[5] whereas urban numbers remained static at about 13%. In terms of absolute numbers, about 850 million rural people compared with approximately 250 million urban dwellers were in this category in 2003. It is also clear that the 2008–2009 financial crisis is impacting the poor substantially due to loss of employment opportunities and family remittances. So it is not difficult to see why there has been increasing concern about improving water supply and sanitation in the last few decades. After all, many of the major public health improvements in the developed world stemmed from the recognition of the importance of clean water supply and adequate sanitation over 150 years ago.

In 2000, as a response to the link between poverty and lack of development, the United Nations developed and unanimously approved the Millennium Development Goals. There are eight goals, ranging from halving extreme poverty to halting the spread of HIV/AIDS and providing universal primary education, all by the target date of 2015. Included in these goals is that of "halving, by 2015, the proportion of people without sustainable access to safe drinking water and basic sanitation." Significant progress has been made since 2000 with access to "safe" drinking water (see Figure 7.1), although statistics collected by the UN focused on access to an improved source of water and did not take water quality specifically into account.

Figure 7.1 Dams come in all shapes and sizes and different modes of construction. This one in the Republic of South Africa is used for drinking water supply with the large pipe taking the water to the treatment plant.

Photo: Colin Chartres

Progress against the sanitation target is a different matter. According to the UNICEF web site, in mid 2009, more than a billion people gained improved sanitation between 1990 and 2002 (see also Figure 7.2). However, the population without coverage declined by only 100 million. In UNICEF's view,

> "...the challenge to provide improved sanitation will be seven times greater in the crucial years leading up to the MDG (Millennium Development Goals) deadline. The population without coverage will need to decrease from 2.6 billion people in 2002 to 1.9 billion in 2015, a total decline of 760 million people. Meeting this target, and reducing rural and urban disparities, will mean providing sanitation services to a billion new urban dwellers and almost 900 million people living in rural communities, where progress has been slower."

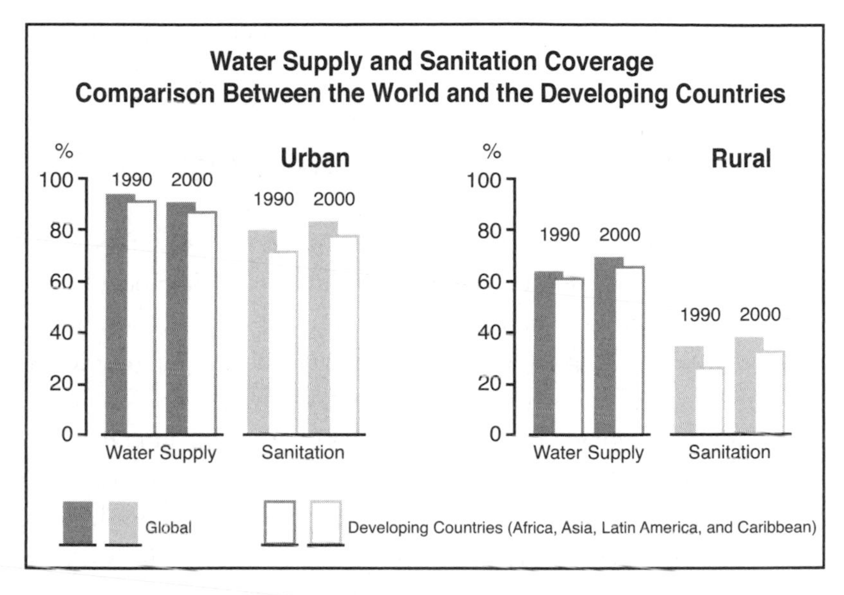

Figure 7.2 Urban and rural access to improved water supply and sanitation

Source: Meeting the MDG Drinking Water and Sanitation Target, World Health Organization (WHO) and United Nations International Children's Emergency Fund (UNICEF), 2006, http://www.grida.no/publications/vg/water2/

The amount of national and international donor support going into the provision of water supply and sanitation (often via non-governmental organizations) is immense. Unfortunately, much of the investment is piecemeal and uncoordinated. Often the basic principles of integrated water resources management that can be applied to conserve the resource and protect the environment are totally overlooked or are only paid lip service. Neglect of water catchment areas mean that we have to pay much more than necessary for clean drinking water due to treatment costs. In other cases, water continues to be used by low-value users, often in agriculture, forcing cities to look for alternative supplies that may include pricey desalination plants and or recycling-treated effluent into the water supply system. Where sanitation is provided at the household level, less attention is given to the treatment of the sewage, and it is often discharged untreated into the environment.

What Happens to Our Waste?

Each one of us produces just under a kilo or about 2 lbs of feces per day and 2–3 liters of urine. Multiplied by almost 7 billion people, this amounts to an unimaginable pile of excrement (almost 7 m metric tons) and lake of urine (about 20 Gl or equivalent of about 20,000 Olympic-sized swimming pools) that are produced every day. When we were predominantly small rural societies and population densities were low, human waste could be readily dealt with in the near environment. In eastern societies, it was and still is used as fertilizer. In Europe, there was an increasing tendency to collect the waste in cesspools or to discharge it into flowing water. However, as societies became more urbanized after the industrial revolution, sewage often began to contaminate drinking water supplies, and following the understanding that contaminated water rather than miasmas (vapors) were the source of cholera in the nineteenth century, many western cities began construction of major sewage systems. However, these sewers discharged the waste still untreated into rivers and oceans. In Europe and North America, sewage treatment gradually became more commonplace. The degree of treatment ranges (see the sidebar at the end of the chapter) from primary, in which floating objects are removed and the remainder discharged to the environment, to secondary and tertiary treatments in which liquids and solids are separated, and the solids are treated by aeration and/or anoxic reaction processes (see Figure 7.3). The resulting sludge is then dried and disposed of in landfills or at sea, is burned, or used as fertilizer. Tertiary treatment involves further processing that may remove nitrogen and phosphorus from the liquid effluent and/or passage through lagoons where reed beds and other biological agents further clean the water before it is released back into the environment. The degree of treatment in the western world varies depending on location of the waste outfall and local, national, and international treaties as well as the degree of public concern about environmental contamination. However, while the technology exists to deal with waste, it is expensive and often beyond the ability of many cities in the developing world both to pay for and maintain. Sunita Narain, a well-known Indian environmental activist summed India's problems up as follows:

> "Currently, the country has installed capacity to treat roughly 18 per cent of the "official" excreta it generates. But it is well accepted that some of these plants do not function because of high recurring costs of—electricity and chemicals or some that do function cannot because they do not have the sewage to treat. This is because, like water pipelines, sewage pipelines will have to be built and then maintained. The fact is that most of our cities, old and new, do not have underground sewerage systems and even if they do, most of the pipes are old and defunct. If all this is put together, then officially the country actually treats 13 per cent of the human excreta it generates. The final blow comes when the partial sewage 'actually' cleaned through expensive treatment gets mixed with the untreated sewage of the majority. This, in a situation, where municipalities are already not recovering the cost of supplying water, forget the treatment of sewage. The result is pollution. The end is the drowning of India in its excreta."[6]

Dealing with problems described here requires a paradigm shift in terms of our thinking and action about how to manage waste. As is shown next, there is some hope that we can do better than we are doing at present, but there are many real problems as well as taboos to overcome.

Figure 7.3 As fresh water resources become scarcer, recycling will become increasingly important. Current technologies involving sand filtration, activated charcoal, ozonation, and pressure membrane micro-filtration (illustrated here) can be used to recycle sewage effluent to drinking water. However, concerns about "toilet to tap" usually mean that wastewater is reused in urban or agricultural irrigation and thus does not require as much expensive treatment.

Photo: Colin Chartres

Recycling and Reuse of Wastewater

Recycling and reuse of waste waters is a complex issue from both physical and social points of view. While debates rage in the developed world about safety and acceptability of wastewater reuse, it is commonplace in the developing work, even if inadvertent. John Radcliffe produced a detailed and comprehensive report on Water Recycling in Australia, which also reviews efforts in other countries.[7]

He points out that recycled water can be produced at area scale (for example, from an urban sewage system), locational scale (from specific sites such as shopping complexes, airports, and so on or from sewer mining for playing field irrigation), and site scale (from individual households). Irrespective of scale and of concerns about whether we should go as far as drinking recycled water, this source of water provides us a great opportunity to deal with water scarcity and an opportunity that must be viewed in an IWRM context.

In several locations, treated sewage is fed back indirectly into the drinking water supply, and in one city, Windhoek in Namibia, it is used to feed a state-of-the-art drinking water treatment plant and goes directly into the city's potable water supply, albeit mixed with reservoir or groundwater. In spite of the "yuck" factor, it is important to point out that there has not been one documented case of illness arising from the reuse of sewage in Windhoek. Elsewhere, communities often balk at the prospect of having even a small quantity of treated effluent fed back into their water supply even if it resides for several months or even years in surface storages after treatment. This attitude, which is often driven by misguided concerns from the medical profession and traditional engineering standards of separating potable water from wastes, is somewhat ridiculous as the majority of the world's towns' and cities' drinking water flows in from upstream, where users dispose of their effluent, treated or untreated. Great faith is put in the ability of a "few magic miles" of river ecosystem services to clean up the water. However, some communities such as those in Orange County, California, are becoming increasingly used to reusing water that has been treated and then injected into groundwater aquifers. While reuse of such water is primarily for irrigation of parks and gardens, there are places where potable reuse is also taking place. We have to become much smarter in terms of not only the investigation of opportunities to reuse wastewater and urban runoff, but in the way that we help communities understand processes of water treatment and the options and potential risks of reusing treated waters for a range of purposes.

In the developing world, the situation is generally somewhat different, given that sewage treatment plants are rare. Effluent is often transported directly into local drains and streams and thus back into rivers and lakes without any treatment. Consequently, problems of contamination from petrochemicals, feces, hospital, and industrial waste including heavy metals are commonplace and impose very significant risks on downstream water users and the environment. Depending on effluent discharge rates, dilution factors, and the ability of the riverine ecosystem (the fauna and flora associated with rivers and floodplains) to "clean up" the polluted water via nutrient removal and uptake into food webs, most developing country rivers are far from pristine and can be likened to what was observed in Europe and the industrial parts of North America in the nineteenth and first half of the twentieth centuries.

However, in many countries, particularly those that are water-scarce or have seasonal rainfall regimes, sewage effluent is often prized as a water and nutrient source by opportunistic farmers. The nutrients including nitrogen and phosphorus are free, and the water transporting them is available year round compared with natural sources. A recent study[8] that surveyed 53 cities across the developing world demonstrated that approximately 200 million farmers and their families depend on untreated or partially treated effluent as a water source. At least 700 million people were considered to consume vegetables that are the predominant crop from such wastewater farming systems, by opportunistic farmers, as shown in Figure 7.4.

Figure 7.4 Urban wastewater near Hyderabad, India. This water is reused for irrigation of Para grass which is used as fodder. There are considerable health risks to farmers and consumers of water from using wastewater and from wastewater irrigated produce. More attention and investment in sewage treatment will be essential in developing countries to minimize these risks.

Photo: Sanjini de Silva

The study showed that out of the 53 cities studied, only 8 reported to have little or no irrigated urban or periurban wastewater irrigated agriculture. Seventy-four percent of the cities studied had wastewater agriculture, though data on extents was not available for some of them. Where data were available (31 cities in this case), cumulative figures show that there are about 1.1 million farmers around these cities making a living from cultivating 0.4 m ha of land irrigated with wastewater (raw or diluted wastewater and includes all those areas that use polluted rivers as the irrigation water source).

Irrigation using wastewater is not a new phenomenon especially in the periurban areas around cities. The periurban fringe, particularly in developing countries, is an area of transition where agriculture is being lost to industry as cities grow and their populations burgeon. There are two fundamental reasons for why wastewater is being used for irrigation—the first being that in the absence of adequate sewerage infrastructure, especially in the outskirts of cities, most of which have grown without much planning, urban waste flows out and mixes in with irrigation water. Farmers in these areas are, then, required to use wastewater for irrigation, which, as was mentioned before, many have grown to prefer. The second important reason is that wastewater is often a more reliable resource in these areas. Irrigation water supply in areas of scarcity and with poor irrigation infrastructure cannot be counted on to provide a reliable and regular supply of water for farmers. The constant discharge of wastewater from the cities, on the other hand, is seen as a more regular source and one with nutrient value.

Efforts to improve the regulation of wastewater use in agriculture at an international level have been spearheaded by the World Health Organization (WHO) and International Water Management Institute (IWMI). Both organizations have been instrumental in defining guidelines for wastewater use to minimize the health impacts of wastewater on farmers who use the resource, as well as for consumers buying food crops that have been grown with wastewater (see Figure 7.5). However, these efforts have been mostly in linking sanitation and health departments together to reduce the incidences of water-borne diseases. Similar linkages at the level of local departments and agencies overseeing sanitation and agricultural water management still need work in order to better reflect the ways in which the different uses of water impact each other. As well as the issue of direct pollution of water courses with urban sewage, animal waste from farms must also be considered (which is generally even larger in volume than sewage) and non point-source pollution coming from agricultural and other developments in water catchment areas. Integrated water resources management is based on the principle that it is necessary to look at the whole water cycle from urban to rural water use and particularly to pay attention to downstream impact. It still has a long way to go in most developing countries.

Figure 7.5 Woman selling vegetables grown with wastewater irrigation in Hyderabad, India

Photo: Sanjini de Silva

Multiple Use Water Systems

One of the biggest obstacles to the achievement of IWRM is the way in which we structure our laws and policies to separate the different uses of water. Sectoral thinking (that is, viewing water resources and their management from a single issue point of view) in water management has been the norm for years, and it is the basis upon which water is allocated for different uses. For instance, water management is usually defined on the basis of its use in the agricultural sector, for industry, sanitation, domestic use, and so on.

In most rural areas of the world, there is no real difference in supply between water used for agriculture and water used for domestic uses such as drinking (as shown in Figure 7.6), washing, bathing, or cooking. Centuries-old tank systems were in fact built with the understanding that the water would be used for multiple purposes and accommodated these uses; for instance, tanks were built with steps on the sides for people to bathe and wash. The limits of sectoral policies with regard to water management are most acutely felt by the poor. Homestead gardening is an important source of additional income for poor families, providing a means of employment for women who are generally responsible for them and also as a source of food security for families in times of difficulty. Studies that have looked at domestic water supply and its use in home gardens have shown that with a small increase in water supply, the additional income that could be gained for the family can help to empower women with a source of livelihood and be a form of food security for a family in times of stress.[9] In fact, the assurance of water supply shows that people will use the resource for both domestic and productive purposes. The limitation of domestic water supply provision on the basis of established norms such as those that calculate and individual's need as 15 liters per day restricts the ability to use water for any other productive purpose. For poor rural households the additional water could help support rearing livestock or producing handicrafts. The implementation of multiple-use water systems that are designed based on socio-economic and physical need principles is one relatively easy way in which we can significantly help the rural and periurban poor and, in particular, poor women.

Figure 7.6 Ensuring safety of drinking water is critical everywhere. Here, simple filter systems are used outside a Nubian house near Aswan, Egypt.

Photo: Colin Chartres

Conclusion

The IWRM approach encourages a change in the way we plan water development. It encourages a view that brings in greater perspective to ensure that the systems put in place reflect that reality of water use. The book's final chapter discusses in detail the experiments underway to improve the planning and management of water through democratic processes that are inclusive and involve a wide range of stakeholders in the decision-making process. Starting with local government, the IWRM approach advocates for dialogue to improve the planning process, and this is to be carried out up to the higher levels. IWRM is not lacking in critics—it is still very difficult to change entrenched views on water resources planning, to get past the sectoral difference that most water planners are used to, and to re-imagine planning.

Stages of Sewage Treatment[10]

Preliminary Treatment

Physical screening of arriving influent to remove coarse particles including stones, sand, and gravel

Primary Treatment

Removal of majority of remaining particulate matter, involving comminution if necessary, coagulation, flocculation, and sedimentation

Filtration can be applied after sedimentation as an alternative. This process removes about half of the suspended solids and reduces the biological oxygen demand (BOD) as well as about 10% of the nitrogen and phosphorus. If primary effluent is discharged, it is sometimes disinfected.

Secondary Treatment

A range of options involving a range of aerobic biological processes aimed at the microbial metabolism of dissolved and suspended organic matter such as aerated lagoons or stabilization ponds

Faster processes such as activated sludge digestion or alternative fixed film processes such as trickle filters may also be used. Some of the organic matter in the wastewater provides energy and nutrients for microbial populations, the remainder being oxidized to carbon dioxide, water, and other end products. A secondary sedimentation is used to remove the biomass produced, and then this can then be treated by aerobic or anaerobic digestion, composting, or other technologies. These processes remove up to half the nitrogen and convert the phosphorus to phosphates. There may also be a further filtration of the effluent stream and disinfection of the effluent before discharge. About 80–95% of the BOD and suspended solids are removed by this stage of treatment.

Tertiary Treatment

Further removal of colloidal and suspended solids by chemical coagulation and filtration with the removal of specific metals, pathogens, and nutrients

Activated carbon may be used to absorb hydrophobic organic compounds and lime to precipitate various cations and heavy metals at high pH level. In modern advances microfiltration and reverse osmosis are being adopted. These result in essentially pure water fit for a range of uses.

chapter eight

Water Governance for People and the Environment

"Whiskey is for drinking; water is for fighting over."
attributed to Mark Twain

Water governance is about effectively implementing a socially acceptable regulation and allocation framework for water and is thus an intensely political process.[1] Governance is a more inclusive concept than government, *per se*; it embraces the relationship between a society and its government and generally involves mediating behavior via values, norms, and often through laws. This complex series of interacting principles and behaviors clearly demonstrates that dealing with water issues is not simply an engineering or science-based process.

Consider the following example: Imagine the task of building an irrigation system with a dam, reservoir, and canal network and then add the challenge of managing water in a country. The first thing needed is money—lots of it. The investment for funding large-scale irrigation schemes most often is sought through a mixture of funding from national governments together with loans from international financing agencies such as the World Bank. The country makes its justification for the large investment on the basis of the economic growth that the additional water resources would support either through agriculture or industry. It is also justified in terms of welfare—that is the number of hungry mouths it would help feed and employment it would generate. The scheme is owned by the government, which means that all decisions regarding construction and quantities of water to be released to various users for diverse uses is determined by a formal structure and system of rules. This would involve state bureaucracy as well as politics and a lot of lobbying. The system, once built, has its own structure for management, and the organizations and agencies within it will be responsible for allocating and releasing water to different groups that have been granted entitlement to the resource. They will also monitor and record the quantities of water in the reservoir and water flowing though the canals and smaller distribution channels that serve farms and fields.

A question arises once the water begins flowing through the canal about how the farmers are to share the water that flows through it and pump it into their fields. The decision of who gets how much water is originally determined by the organization's rules. However, in a particular village state agents have been lax about monitoring quantities for the farmers and have also given in to corruption. A few powerful farmers have been able to take more water and also have bribed the state agents to release water at times outside of the approved schedules. In the past there had been a customary arrangement that determined how the farmers shared water, which was based on the traditional rules of the community. This eventually broke down once the state set out its rules, and the farmers are unhappy with the situation and feel stuck. Meanwhile, their livelihoods have suffered.

Consider now, that the river upon which the original dam was built flowed through two countries. We have just been talking about country one, where the river originates. Country two is much poorer than country one, in large part because of country one's ambitious plans to increase agricultural production, which has resulted in irrigation development that has taken water away. By the time the river flows into country two, it is a mere trickle in the summer season and a little more than a large stream during the rainy season. Many years before, both countries signed a treaty that determined how much water would go to each, and country one's plans to dam the river were approved as it was assumed that there would still be enough water to meet country two's needs. However, climate change has greatly affected water availability and has contributed to the declining amount of water in the river.

These scenarios exemplify not only the interdependence of water governance on climatic factors, but also that transboundary issues, allocation policy, corruption, and farmer needs all have to be taken into account in planning and managing water resources. Governance is the process by which decisions are made within a framework of rules, laws, and policies. These decisions therefore would typically involve a wide range of individuals and organizations from planners and practitioners at the municipal, state and central level, to various interest groups, companies and corporations, and communities. Simply understanding the various aspects of governance and the uniqueness of water resources makes any linear thinking on its management impossible. To consider the competing users and uses of water together while also ensuring that the resource is managed efficiently and allocated equitably, governance frameworks must be able to balance the different levels at which water is managed in a formal, yet flexible way. The success with which we meet the demand for water to support the activities crucial to our sustenance depends on this framework.

Water management systems have evolved greatly in the past 50 years from the predominance of centrally planned and controlled systems. Farmer-managed and community-based systems are becoming increasingly common, as are decentralized governance models where local government authorities along with communities play a larger role in water management decisions. The issue of privatizing water is increasingly debated as a more efficient alternative, and despite its controversy over legitimate concerns that the poor will be exploited by the profit-making drive of large companies, it makes for a compelling case. In many countries, it is the poor and the marginalized who have no connections to public utilities that rely on informal water markets to meet their needs. Issues such as this also bring into question what the new role of the government and public sector should be in the wake of new challenges, such as unprecedented population growth and climate change.

Drawing on experiences from different countries on what has worked, what hasn't, and in what context, the chapter touches on the principles of good water governance. It helps shed light on what the options are for private, public, and community management based on lessons learned from the successes and failures of these approaches. Did they help to reduce poverty? Were these management approaches efficient? Were they sustainable? And from the failures—what do we now know about doing things differently? The governance of water is impossible to extract from the governance of other sectors, particularly agriculture and increasingly energy. We also discuss how policies and decisions in sectors that are influencing water policy are playing a role in determining new complex governance mechanisms.

Understanding Water's Unique Properties

The basis of the water governance challenge lies in the many ways the resource is accessed. For most of us, water is assumed to be either free or provided for by the state or private supplier for a fee. Water is "free for all" when it is accessed freely through *open access* from a communal source such as a river or lake; it is a form of *common property* when a formal or customary law regulates how and whom might use it; it is also linked to land ownership when it is pumped or piped from the ground or a water source on land that is privately owned, in which case it is a *private good*; and finally it can supplied as a public utility such as a municipal council, in which case it is a *public good* owned by the state.[2]

The fact that water can be accessed by more than one of these ways simultaneously is what makes the governance of the resource such a challenge. Moreover, each of these different ways of accessing water resources has socio-political, economic, and technical dimensions. Each of these dimensions could also affect the rules that determine when and how water can be extracted and who can use it, as well as the cost or rules of exchange that determine its use. Conditions such as these would then determine whether water is a private or a public good. For instance, take the farmer with a pump who can draw water from a reservoir: The water in the reservoir is a public good paid for by public investment in water storage infrastructure, but it then becomes a private good when on the farmer's property. Similarly, the technologies used to control the resource can also determine whether water is a private or public good. For instance, the farmer with a groundwater pump has control over water because of possessing the ability to draw it from an aquifer, thereby making it a private good. However the aquifer itself could also be a public good, as the extent of groundwater is difficult to measure and could extend well beyond the farmer's land. Water is also a *common pool resource*, which means that the complexity of its state and existence in different forms, for instance as groundwater or surface water, and its ability to move across private, political, and geographical boundaries, make it impossible to exclude its various uses and users who are competing for the resource.[3]

All of the modes by which water is accessed are part of formal or informal institutions that provide rules, laws, and codes that determine how it can be used. Formal institutions, such as the laws of the country, are those that have a framework of rules, policies, and organizations that together implement and monitor those laws. Informal institutions are a little trickier to identify as they are not generally documented or protected by a network of policies and organizations. Informal institutions are often linked to customary or traditional laws, such as in the case of a community that has practiced agriculture using the water from a wetland. The community's traditions may include rules for how much water can be drawn from the wetland, the time of year during which it is allowed, and perhaps even how responsibilities for cultivation and irrigation are shared among community members. The community would then be the protector and enforcer of these rules, which may or may not be known in the wider region or state as they are based on the trust and relationships of the members of the community. In many cases governments have assumed a water source to be open access when there had in fact had been informal arrangements for its management that were created and overseen by the communities around it.

Water governance functions at different levels—local, regional and global—to address the issues of management at different scales, the field, the farm, and the river basin. Starting with local governance and management, let's look at how thinking has evolved and the challenges that have shaped what we now know about water management.

A New Blueprint

The technological challenge to physically move water has preoccupied engineers for thousands of years but never so much as in the last century (see Figure 8.1). The feats of human engineering devised to overcome natural obstacles and deliver water to people also included the planning and construction of reservoirs for storage and vast networks of canal and piped systems. Together these developments have enabled us to bring water to huge areas of previously dry land and have allowed food production to expand to feed growing populations. In the new nations that emerged mid-century, the focus on self-reliance in food production was a symbol of independence and freedom from colonization. In Egypt the Aswan dam, despite its idea being years in the making, became a symbol of nationalism and the then President Nasser's determination to modernize. Another example that is cited often is the inauguration of the Nangal Canal of the giant Bhakra Nangal dam in 1954, India's first hydropower scheme. Nehru proclaimed it a holy place, one he revered more than any other place of worship as it stood as a testament to the lives that were given for the betterment of the lives of others.[4] Food production was the primary objective of these endeavors in developing countries, to ensure that the ravaging famines and gross inequalities of the previous era would not be experienced again. Centrally planned systems that were owned and managed by the government were seen as the most appropriate way in which to plan and administer the delivery of water for agriculture.

Figure 8.1 Water is heavy and requires considerable amounts of energy to move it from place to place when gravity cannot be used. Here a pumping station on the El Salam Canal in Egypt carries Nile water into the Sinai Desert where it is used for irrigation.

Photo: Colin Chartres

Toward the second half of the century, the technological focus was replaced by a new concern when it was observed that despite millions poured into the construction and development of water infrastructure, poverty and hunger still had not been eradicated. Large groups of people, particularly in the rural areas of the world, had been left out, deprived of the transformation that irrigation promised. Coupled with rising population rates, this meant that the problem was only going to get worse. It was at this time that experiments with changing the way water was managed began—essentially with the quest to find alternatives to centrally managed irrigation, owing to its poor performance. The shift in thinking was brought on not only by the need to connect more farmers to canal systems to improve their farm productivity, but also to improve the general efficiency of managing large, public schemes that had become unwieldy and bureaucratic. It was clear that the management of water infrastructure required reform in order to get more crop per drop, but there was also a clearer recognition that people needed to be put at the center of water management. New models need to better reflect the priorities of the farmers, the needs of the poor whose livelihoods depend on agriculture, as well as the human resources and the skills and expertise of those working within the systems.

Going Local

At the massive Earth Summit in Rio in 1992, it was decided that the governance of water needed to be decentralized in order to improve efficiency and equity. The subsidiary principle posits that all matters that can be handled at the local level or by the smallest organizational unit should be handed over to them. This was the overarching idea behind the decision that resulted in the recommendation that farmers were to have a greater role in the management of irrigation systems. Irrigation Management Transfer (IMT)—the policy and process by which irrigation management is devolved from agencies to the farmers directly—took off in the early 1990s. Many countries quickly bought into the trend to decentralize, having experienced less than average performance from their irrigation schemes for some time. The primary motive behind it was reform of the irrigation sector through the IMT policy model, which would transfer responsibilities for management to the farmers. This would include decision-making power on water allocation and responsibilities for operating and maintaining the canals, and it was also seen as a way to lessen the huge financial costs of irrigation infrastructure upkeep. By transferring more responsibilities to the farmers, the overall costs of maintenance would come down—a big bonus for public schemes where maintenance counts for one of the largest costs incurred in irrigation management. With farmer involvement, the incentives for upkeep are built-in, while participation ensures that *their* priorities are accounted for in the decisions that are made about water use and allocation.

The principles of community management and local governance greatly influenced the experiments with decentralization that evolved. The most widely supported, best-practice, approach for running devolved irrigation management institutions is called *participatory irrigation management* (PIM). The smallest unit of PIM is the Water User's Association (WUA), a formal organization that comprises a group of farmers who meet to make decisions regarding how much water they required, what maintenance work needs to be conducted on the canals, how the work could be shared, and what investments need to be planned for upkeep of the system in the future. A WUA can be formed around canal systems, tanks, rivers, or large public irrigation schemes and would also serve as a forum to manage conflicts that may arise between farmers or other water users in the area. All decision making within the WUA is to occur in a participatory way to make the process equitable and inclusive.

The appeal of IMT and PIM can be seen in their spread over the last few decades. Many countries such as Mexico, Argentina, Colombia, Chile, Egypt, Morocco, and Turkey have adopted IMT and PIM. It has also been especially popular in Asia, and since the 1980s India, Pakistan, Philippines, Vietnam, Cambodia, Indonesia, Nepal, Thailand, and Sri Lanka have adopted policies to make the shift to PIM. In many parts of Asia IMT and PIM dominate irrigation reform and development in the sector, and several new models have evolved based on the institutional context in each of these countries. In many cases, the shift to IMT/PIM was championed and financed by international lenders and donors such as the World Bank and the Asian Development Bank. Although proponents of IMT often pointed out that there was no blueprint to implementing the policy model, only guidelines that could suggest what good practices might achieve, many countries adopted the model with backing from international financial institutions, which placed conditions on the way in which IMT was to be implemented.

The Farmers Will Do It for Themselves

About 20 years ago, Mexico began a process to transfer irrigation management, starting with a policy to create water user's associations. It was initiated to reduce the public expense on irrigation management and address the low performance and low productivity of agriculture, recognizing that the central departments did not have the funds to allocate to irrigation operation and management. Ten years later, the program was widely hailed as a success due to the short duration within which it was rapidly implemented across the country.[5] IMT was rolled out first by creating a central water authority called the National Water Association (CNA) that was responsible for the transfer. The authority reorganized all 82 irrigation districts into new units, each with its own WUA. This enabled the WUAs to function with autonomy within their regions, and they were meant to report back to the water authority. The success of Mexico's model was evident from the significant costs that the CNA was able to recover; the government had covered over 80% of operation and maintenance costs during the 1980s, and about ten years later it was reduced to about 25%. Farmers surveyed around that time expressed that the system had improved since the devolution of power. A second generation of problems faced by Mexico's irrigation management appeared around 2000 when it was observed that farmers were paying about four times more for water than they previously paid. The infrastructure was in decline, and as competition from the cities has increased, so have conflicts over water between the municipalities and farmers. The WUAs are also suffering from a lack of human resources, and nepotism and corruption has been noted to occur within the associations.[6]

In the Philippines, IMT was initiated in the 1970s, prompted by financial reasons, and the devolution of power was planned to occur in stages as part of a longer strategy for irrigation reform. The National Irrigation Administration (NIA), established in 1964, was experiencing difficulties in fulfilling its mandate, and it was unable to attend to operations and maintenance concerns. Partial reform was initiated based on the idea that a slower process would help to instill learning at each stage. The system was also based on a centuries-old practice of community management, where small irrigation systems were built and run by the communities that lived around them and the cultural value of *bayanihan*—the sharing of work and responsibilities practiced within its agricultural communities. By using community organizers that went around villages and organized farmers, Irrigation Agencies (IA) were formed, which then entered into formal agreements with the NIA for irrigation management. The IAs were supported by a unit especially formed to improve their skills in management through training. The most promising stories that have emerged from the Philippines' experience so far is that fee collection is most efficient in areas where responsibilities were most devolved, indicating that a greater level of autonomy for the IAs has proved to make them more effective. However, because IA membership is voluntary, staff numbers have reduced over the years, and there are issues concerning the dependence of IAs on the NIA.

History shows that leaving specific groups out of decision-making spaces due to their social or economic status has proved not just to be politically dangerous, but also not economically viable. IMT policy helped to raise the issue of women in water management as many women farmers were left out of WUAs or farmer groups as it was assumed they only perform certain types of work. In agriculture, the assumption that all farmers are male resulted in decades of development assistance and initiatives being targeted to men to transfer technical knowledge and power that left women completely out of the system. With participatory management, all members have a voice, which means that levels of poverty and gender do not exclude people from planning and allocation of water. In countries where women play a major role in agricultural activities, the participatory principle of WUAs is designed to ensure that the membership reflects women's representation. Thus, PIM is believed to enable vulnerable groups to be able to voice their concerns and play an equal role in decision making.

The Participation Principle

Wholesale adoption of the "ideal model" of IMT has occurred in several countries and continues to until today. Although there is still widespread debate as to the success and validity of IMT policy, few can argue with the logic that with a participatory system where the beneficiaries of irrigation can make decisions for themselves with autonomy, a forum to discuss conflicts and also powers to influence the larger decisions is a bad thing. Critics of irrigation reform have also pointed out that too much emphasis has been placed on reducing the financial burdens of the state and not enough on alleviating poverty and empowering farmers. Critics also point out that IMT's premise is the failure of public irrigation systems to be efficient agents. Some also say that it assumes that the government wants to hand over the power of resource management to the farmers and assumes that farmers want to have the power and the added responsibility. IMT was thus promoted initially on a pessimistic note. For farmers who are not used to paying fees for irrigation water, the initial costs that they have to bear in addition to the labor they supply for maintenance of the irrigation canals is high. This also affects the extent to which farmers buy in to the idea of IMT.

Another issue that is often raised is of the technical skills of WUAs. In most IMT policies, technical training has been an important aspect of the transfer of responsibilities; however, this adds to the initial costs of a transfer program, which is often neglected in implementation of devolution.

The fact remains that most of the world's rural poor depends on farming. IMT requires learning and change for both farmers as well as irrigation agents. It is clear that there are certain principles that can be applied to good practice in transferring irrigation management, which include recognition of the rights to water and goes along with acknowledgment of the water service. Responsibilities and roles are clear to those within the managing authority, and there are clear incentives and bonuses attached to doing the job well. In terms of good governance, there are rules and processes to resolve conflict between water users. Perhaps what is clear from the experiences of IMT is that the government cannot entirely remove itself from its role to ensure that laws are being followed and that communities are not being marginalized or discriminated. While having a clear legal institutional framework that includes delineated rights to water is important, these can only be implemented effectively if regulation also stipulates how and what the role of the government in irrigation management should be in the monitoring, guidance, and supervision of irrigation.

Recognizing Informal Institutions—The Commons and Community Management

One of the criticisms leveled at IMT is that in implementation it has sometimes assumed that no preexisting arrangements to manage water existed. Many countries in the world have centuries-old irrigation systems where communities have been practicing agriculture and managing their water themselves. The legacy of these systems still persists today, and the imposition of a state's view of how water users should organize overrides the customary laws or traditions that still live. In some cases, such as in the Philippines, IMT was based around a communal arrangement for water management that was rooted in certain cultural values, and so communities that had experience managing resources and sharing responsibilities were slowly introduced to the formal irrigation association. These assumptions have destroyed many existing common property regimes. Efforts to bring indigenous institutions back into the mainstream of water management have required extensive anthropological investigation as these institutions have slowly eroded over time. Rules imposed from colonial eras, the modern nation state, and the fragmenting of societies caused by economic and socio-political factors such as wars have all contributed to the wearing down of these systems.

One of the most interesting examples of indigenous community-based institutions that are being changed is in the formalizing of water rights. In South Africa, the state has initiated a program to issue permits to water users. It follows on from a trend to formalize water rights in various countries and has faced some opposition from community groups who claim that it overrides pre-existing governance structures. In South Africa the permit systems have been put in place with equity in mind—the problem with a lot of community based organizations is that local power struggles between groups often dictate who does and does not have rights and access to water.[7] By formalizing the system the government can issue licenses and monitor the use of water to ensure that everyone has an equal right. However, there is also value in recognizing indigenous or community-based institutions as very often they are potentially more sustainable and long-lasting as they are built into community values and may have a long history.

Some say that formalization is not as suited to developing country contexts where it is difficult to regulate and monitor rules and where corruption and local power dynamics mean that is it more likely that the rules are overridden by a more powerful group depriving the rights of the poor or marginalized. A more dynamic set of laws that brings together formal and informal aspects would perhaps work better as it ensures that there is recognition of local customs and traditions, not forgetting the livelihoods around which the rules were originally established.

Rivers

In most countries, water resources are allocated for different uses. Water for drinking and household use and, initially, agriculture are given top priority, followed by water for industry and for generating electricity. As the competition for scare resources increases, the potential for conflict between various users and uses of water also increases. Conflicts over water have even been speculated to lead to wars in the future. Up until now we have talked about how the most efficient and sustainable ways to manage water are at the local level. However, as the issues surrounding water scarcity and access are becoming more serious and given the nature of water and the fact that the rivers from which most of the fresh water supply of the world depend traverse political and geographical borders, the river basin is becoming one of the most important scales at which water is managed. The river basin is logically the most appropriate scale at which water should be managed, as it defines hydrological boundaries. Unfortunately hydrological boundaries and political boundaries rarely match up; if they did it would make water management a lot simpler. River basin organizations were initially set up to address how planning for water resources within countries could be aligned given the competition for water resources within the basin. In addition to being a way to resolve conflict and raise issues concerning the equitable sharing of water resources, river basin institutions are also seen as a way to regulate water use through treaties and laws or monitoring bodies.

The need for an institutional structure that addresses water allocation at the level of the river basin is widely accepted. River basin institutions such as the Mekong River Commission (MRC) are working to try to harmonize the interests of Laos, Cambodia, Thailand, and Vietnam, as the approximately 60 million people living along its banks depend on the river for their survival. However, the impediment to some of these regional organizations, such as the MRC, is making sure their efforts actually work to support national priorities and policies. The efficacy of such institutions is closely tied to this, and often the problem for these institutions is the extent to which they are seen as impact and action-oriented decision makers.

River basin institutions are also an important place for environmental issues to be raised as the degradation of ecosystem services (which are the products of the environment that we have a demand for as humans) affects us all. By creating a broader platform at which environment issues can be regulated on or policies could be constructed to protect biodiversity of the region. Although in many cases, regulation within river basin organizations aren't very powerful, and without strong enforcement, the forum still provides an important space for discussion and information sharing on transboundary water.

Conclusion

Perhaps the single most important principle of water governance is that institutional pluralism is required—a system that can accommodate a diversity of approaches and be flexible enough to change and adapt to new circumstances.[8] The very nature of water makes it impossible to effectively govern it by a single system. We must acknowledge that a range of institutions, formal and informal are operating at a range of scales: local, national, and global. Those who take effort to reform the governance of water will have to be aware of the fact that the effect of changing any institution or policy will be felt through all of the others. It is also becoming evident that because of the links that water management has with other aspects such as energy and agricultural policy, it is very difficult to make decisions about water without looking at the big picture.

chapter nine

Water Rights and Water Costs

"When the well is dry, we know the worth of water."

Benjamin Franklin, 1746

Many of us think little about buying a 500 ml bottle of water at a cost of maybe $1. However, in reality this is more expensive than gasoline. Translated into liters and gallons it would be $1 million per thousand liters (264 gallons). In contrast, we balk at paying higher prices for our domestic tap water. In Canberra, Australia, for example, there is public upset when the domestic water tariff goes over Aus$3.70 (US $3) per thousand liters. Furthermore, this water is perfectly fit to drink. The concern over paying for water arises because people have grown up believing that access to water, at minimal cost, is essentially a basic human right. While few people would argue that access to a minimum amount of water for drinking, cooking, and sanitation to meet basic needs is a right, this right has also been extended to major users of water, such as agriculture and industry. Although water does fall from the sky as a free good, it can be a very costly exercise to build the necessary infrastructure (dams, canals, pipelines, pumping stations, boreholes, water purification plants, and so on) to deliver it to users and to maintain such infrastructure for years to come. The problem is that we do not really know how to value water, and in fact, there is little relationship between the actual domestic price of water and its scarcity.[1] Further compounding the complexity of this is the fact that there is considerable agreement that the environment should also have a defined allocation of free water in order to conserve and maintain the essential ecosystem services that we depend on. Unfortunately, the environment usually gets left with what is not used by others.

While systems of water governance have evolved in most western countries (see Chapter 8, "Water Governance for People and the Environment") to enshrine in legislation rights to access and use of water, the situation in developing countries is variable. Most countries have a plethora of policies, laws, and regulations controlling water. However, access to water is still a major issue for many poor people. This is because the institutional framework, to ensure access and rights do not always function. There are also great debates over how water is to be valued and rights are to be determined, made more complex by issues such as customary and traditional rights, land tenure, the vulnerability of marginalized groups of people, the scarcity of resources and, of course, power and politics.

This chapter explores some of the background behind water rights, valuing, and pricing water and whether or not there are benefits to be had from trading water among users. As these legal and economic dimensions of water management are being recognized, the questions we also need to ask ourselves are: What does this mean for poor communities? Is it necessary to have regulation on every aspect of water management? Who do these regulations benefit?

To reiterate what has been said in earlier chapters, water is becoming increasingly scarce in many countries as populations grow, diets become more sophisticated, competition between water users grows, and climate change starts to bite. This means that we cannot continue to consider water as an almost infinite renewable resource, and that on an annual or seasonal basis, at least, we have to view it as a finite resource. However, we cannot simply view water as a commodity like coal, oil, or wheat because of its fundamental life-giving significance to all of us. Viewing water as something that can be simply treated with a microeconomics approach and sold to the highest bidder will simply not work. Rather, we do have to consider how much water should be a basic human right and how access to water for all people can be protected. We also need to look at procedures for allocating available water to different sectors of the economy and to the environment and the nature of our institutions and governance policies that facilitate this. Once we have such processes in place, microeconomic approaches will be extremely useful to differentiate between optimal and less optimal uses of water and ways of saving water in water-scarce environments.

As a starting point, this chapter looks at how these factors are dealt with in a number of developing and developed countries to assist in the definition of more universally applicable principles.

Water Allocation Mechanisms

The dangers of open access to and subsequent degradation of the "commons" was first talked about in Gareth Hardin's essay, "The Tragedy of the Commons."[2] Though his analysis has now been questioned, Hardin's statement about the ways in which rapidly reproducing human beings seek to maximize their advantage from the natural world of finite resources and its frightening consequences led many to seriously consider what could be done to protect natural resources. In 1992, the influential Dublin Principles[3] on water recommended that "managing water as an economic good is an important way of achieving efficient and equitable use, and of encouraging conservation and protection of water resources." Many saw it as a push to privatize water. Subsequent attempts by the private sector to take over city water supply and sanitation networks have only been partially successful, primarily in developed countries. Major problems with such privatization have occurred in several developing cities (Manila and Buenos Aires are examples), including financial viability, the cost of replacing aging infrastructure and poor people's unwillingness, and sometimes inability to pay for the service.

If we are to progress toward more equitable sharing of water resources and if the market is to be allowed to eventually deal with scarcity via pricing mechanisms, then the first step on this journey has to be focused on the definition of water rights. The following sections examine what has been achieved in this regard in a number of countries.

South African Water Policy

At the end of the apartheid era, South Africa ambitiously decided to reform its water laws as a response to both perceived injustice in access and allocation and to growing water scarcity in what is essentially a semi-arid country. (43% of the rainfall falls on just 13% of the land.) When apartheid crumbled in 1994, an estimated 14 million South Africans lacked access to a formal water supply, and about half the country (21 million people) had no formal sanitation, according to the Department of Water Affairs and Forestry.[4] Since then, access to water has increased dramatically, but backlogs persist: In 2008, about 5 million people were still in need of adequate supplies, while three times more—15 million people—lacked basic sanitation according to the UN.

The following challenging goals are fundamental to the South African Water Act:

- Abolishment of the rights previously granted to landowners, who were almost exclusively white, to ground and surface water on their property.
- Making the central government the public trustee of the nation's water resources.
- Ensuring access of all South Africans to enough water to meet their basic human needs and recognize aquatic ecosystems as having a legitimate right to their own water. Water

for minimum human and ecological needs constitutes an untouchable "reserve."

- Honoring water provisions contained in international agreements and treaties.
- Economic pricing of water.
- Establishment of water management and user agencies at the regional level within hydrological areas known as catchments.
- Issuing of licenses for water use and review of licenses every five years.
- Imposition of water-use charges for discharging pollution directly into the resource.

From the perspective of this chapter, the key issues that the South African Water Act highlights are the recognition of the right for all South Africans to have enough water for their basic human needs and the inclusion of this water in an untouchable reserve along with environmental water. Additionally, divorcing water rights from land rights and the recognition of water catchments as fundamental units of water management, the establishment of user agencies and licensing do put in place some of the basic prerequisites if water is going to be eventually priced and equitably shared.

In terms of pricing in South Africa, the price of water sold from government water works is set each year in April. Water is seriously undervalued in South Africa. This tends to mitigate against water being viewed as an economic good and hence being valued and conserved as it should be in a semi-arid country. This also indicates that the price of water must reflect its cost, but it is also necessary to recognize that water is not only an economic commodity but also a social commodity closely related to other factors such as health and production. The price of water must therefore reflect the difficult tension between equity and economic sustainability.

Establishing a governance framework that ultimately enables more rational water allocation and pricing policies is no easy task. It has been compared to planning an expedition up an uncharted mountain. The authors of a recent report identified three separate tiers that require negotiation:[5]

- The broadest tier, at the policy level, which may be national or regional policy
- The second tier, at the strategy level, which may be at water management area or catchment scale
- The third tier, at the local operational level within a catchment

They also indicate that the time to achieve the desired outcomes may vary. The changing of policy at the level of national constitutions may take 50 to 100 years; response to national sectoral (the different sectors of the economy) policy, such as water policy, may operate on time scales of 20 to 30 years; strategies at water management or catchment scales can typically take 5 to 10 years to move through the cycle, while everyday operation at the local level can be adapted within a matter of weeks if necessary.

Water Allocation and Pricing in the United States

East of the Mississippi, most states base water law on *riparian rights*. The essence of riparian rights as the basis for water allocation is that a person who owns land on, alongside, or crossed by a natural watercourse has a legal right to access and use the water running through the property. Note that this is quite different from the new situation in South Africa and, as we will see, in Australia.

According to the National Science and Technology Center (NTSC) of the Bureau of Land Management, the doctrine of riparian rights in the United States has its basis in case law that first involved Tyler v. Wilkinson in 1827.[6] This case arose out of a dispute between mill owners over the right to use the flow of a river for mill power. The opinion in the case stated that all land owners adjacent to a river had equal rights to the water in the river and that an upper proprietor could not diminish the quantity that would naturally flow to the lower proprietor. However, the case opinion also recognized that such an absolute right would not be practical and held that an upper proprietor could make "reasonable use" of the water, including consumptive withdrawals.

Riparian water rights, therefore, linked water rights with land ownership. This water right, however, is described as a usufructuary right (a right that allows the use of property that belongs to another) and not a property right in the water. This means that the water may be used as it passes through the property of the land owner, but it cannot be unreasonably detained or diverted, and it must be returned to the stream from which it was obtained. The concept of "reasonable use" permeates the law of riparian water rights. Reasonable use enables the consumptive use of water, but what actually constitutes reasonable use differs from state to state and is a subject to continued change.

Riparian rights are specific to watercourses and do not apply, however, to diffuse surface waters including storm or flood drainage. U.S. law theoretically views riparian water rights as unable to be lost through nonuse and is indefinite in duration. Therefore, a riparian landowner does not lose their riparian right by not using water. However, the courts tend to provide greater protection for existing uses than for potential future uses. Riparian rights can, however, be lost through prescription. Prescription is a process of involuntary transfer from one party to another. Under prescription, a party making open use of water for the proper time period (usually 20 years) gains title to the water right superior to that of the original holder.

An important aspect of riparian water rights is that they are generally considered "part and parcel" of the land and are included if the property is sold. The law, in most cases, forbids transfers of riparian rights for use on non-riparian lands. This rule, however, has been amended in some instances to allow non-riparians to use the water so long as the use is "reasonable" with regards to other riparians.[7]

In the western United States, *prior appropriation* systems tend to prevail. In a prior appropriation system the first users to withdraw water from a source for a beneficial use have the right to the same amount of water in perpetuity, as long as they continue to use the water for the same purpose. Subsequent users have to respect this seniority under this method, as new users are only allowed to use the remaining water as long as it does not affect the previous users. Allocations are determined annually based on the capacity of the shared water source. The appropriators then receive their full appropriation of water in order of seniority based on the priority date until the capacity of the source is exhausted. When water becomes scarce, junior appropriators may find themselves without any allotment of water. Water rights from prior appropriation are not connected to land ownership and therefore are transferable between users, which theoretically allows for water trading.

It is, however, a difficult business to determine who has what priority. In New Mexico, for example, surface waters must be allocated through a formal lawsuit brought by the state in which evidence of past use is introduced and adjudicated. There are cases that have taken decades and millions of dollars in legal fees to resolve.

There are key differences between riparian and appropriative water rights. An appropriative right is usually based on physical control and beneficial use of the water. They give an entitlement to a specific amount of water, for a specified use, at a specific location, with a definite date of priority. An appropriative right depends upon the continued use of the water and may be lost through nonuse. Unlike riparian rights, these rights can generally be sold or transferred.

In the western states, there are few restrictions on who can hold appropriative water rights. Therefore, both private and public entities hold rights. An appropriative right does not depend on land ownership, but some states do require that the water is linked to the land on which it is used. In general, appropriative water rights are transferable property. There are, however, three major requirements that inhibit the transfer of an appropriative water right:

1. Rules prohibiting the severance of water right from the land on which the water is appurtenant to
2. Showing that there will be no injury to other appropriators
3. Establishing the extent of the water right for transfer

The traditional means of losing appropriative water rights are nonuse or abandonment. Loss through abandonment is a consequence of the essential role that "use" plays in the definition of the right. The right does not come into existence without application of water to beneficial use and cannot continue to exist without the continuance of beneficial use. Nonuse in itself, however, does not always constitute abandonment. A finding of abandonment often requires a determination of an intent to relinquish the water right. A statutorily specified period of nonuse can, in most states, serve as proof of intent to abandon. In other words, an appropriative right can be lost through nonuse when intent to abandon can be demonstrated or when the water right has not been used for a specified number of years.

In some U.S. jurisdictions a hybrid doctrine that recognizes both riparian and appropriative water rights holds sway. Hybrids have evolved because a state might have originally had a riparian rights systems but subsequently changed to an appropriative system. The NTSC Bureau of Land Management indicates that

"Hybrid states have integrated riparian rights into the doctrine of prior appropriation by converting riparian rights to appropriative rights. Generally, states have allowed riparian land owners to claim a water right by a certain time and incorporate it into the state's prior appropriation system. The riparian rights tend to be superior to the prior appropriative rights even if the water was not put to beneficial use until much later. Riparian rights are not recognized, however, if they are not claimed by a certain date (usually the date the state adopted the prior appropriation doctrine), or are not put to use within a certain number of years. States that have a hybrid system include California, Kansas, Nebraska, North and South Dakota, Oklahoma, Oregon, Texas, and Washington."

The Murray-Darling Basin, Australia

In Southern Australia and particularly the Murray-Darling Basin, growing water scarcity, probably due to climate change and increasing public concern about environmental damage due to the effects of perceived overuse of water resources by agriculture, have seen a remarkable and relatively rapid process of reform.

A report prepared in 1999 by the Australian Academy of Technological and Engineering Sciences indicated that if current patterns of water use continue, the water needs of industries will outstrip water availability by 2020–2021.[8] Irrigated agriculture, Australia's major water-using industry, would be seriously affected by the shortfall. The findings of this report have subsequently become all the more prescient because of the severe drought that has gripped Australia's southeast since the turn of the century.

Water rights and allocation processes differ to some extent between the Murray-Darling Basin states and territory (Queensland, NSW, Victoria, South Australia, and ACT). However, historically, water in Australia was obtained by a potential user applying to a state water resources agency for a license to extract water. Under this system, governments theoretically had the ability to manage the resource in a sustainable fashion and keep water aside for environmental purposes. However, in most jurisdictions there has been rampant over allocation of water through issuance of too many licenses for too much water. By the end of the century, over allocation was a particular problem in several of the major rivers of the Murray-Darling Basin including the Namoi and Murrumbidgee.[9] Mitigating this to some extent was that many of the licenses, referred to as "sleeper" and "dozer" licenses, were not actually being used to extract water. The over allocation of water in most jurisdictions did result in efforts by state governments to reallocate or to encourage the reallocation of water to nonconsumptive (environmental) uses, but in general, this was too little and too late.

As an initial response to growing concerns about water availability, in 2003 the Council of Australian Governments signed the National Water Initiative to be supported by a National Water Commission. The National Water Initiative remains the blueprint for water reform in Australia and has at its core

- A water market and trading scheme for the Murray-Darling Basin
- Best practice water pricing
- Working with all parties in the Basin to ensure a balance is kept between all water users, including water to keep wetlands and river systems healthy
- Keeping up-to-date records of our water availability and use

In most Australian jurisdictions, planning is undertaken by the water resources agency to allocate water between consumptive and nonconsumptive uses, based on an assessment of economic, social and environmental benefits and costs. Distinctions are made between a water access entitlement and actual allocations, which depend on how much water is available in the dams and storages at the beginning of the irrigation season. The prolonged drought in southern Australia, however, has meant that many people with water access entitlements were receiving very little of this as an allocation even for what is termed "high security water." This lack of water has lead to severe economic problems for farmers dependent on perennial irrigated crops such as grapes and fruit trees. Similarly, areas of high environmental value were observed to be suffering due to low river flows. These economic and environmental impacts lead to policy responses that saw the relatively rapid separation of land and water rights and the development of both temporary and permanent water trading. The amount of trading out of state is, however, capped on an annual basis by some jurisdictions. To support water trading, a water market has developed with online or phone connectivity (see Table 9.1). Temporary trades between September 2008 and June 2009 were averaging between Aus$300–$400 per ML (264,000 gallons) and permanent and temporary trading continues to grow. It is now viewed by many farmers as an important way of ensuring that they can maintain income in dry years. A central tenet of water reform in Australia was to establish property rights to facilitate trade, allowing water to be redirected to its highest value use and the market allows this to happen.[10] However, while this is an important aspect of water management attention must also be given to environmental water rights.

Table 9.1 *Water Trading by Irrigation Farms, Murray-Darling Basin, 2006–2007 (after ABARE, 2009)*

Percentage of Farms Trading	Dairy	Broadacre	Horticulture	Murray-Darling Basin
Permanent entitlements	3%	1%	4%	2%
Temporary irrigation water	31%	20%	23%	23%

While the National Water Initiative was agreed and endorsed by the Council of Australian Governments, implementation of reform was primarily at the state government level. Subsequently, because of increasing concern over water scarcity due to drought and/or climate change, the Federal Government has had to step in and deal with the over-allocation issue in the Murray-Darling Basin. It has done this through an ambitious Water Act passed in 2007 and amended in 2008 (see sidebars at the end of this chapter). These acts further strengthen the goals of the National Water Initiative. They also, backed up by very significant operational funding, enabled a new Environmental Water Manager to purchase water for the environment, allowed for significant investment to improve system and in-field water efficiency and put in place a water measurement program that will in the future provide the essential scientific information upon which to base policy and management decisions. The same period saw the demise of the Murray-Darling Basin Commission (MDBC) and its replacement with a Murray-Darling Basin Authority responsible for planning in the basin and reporting directly to the Federal Minister responsible for water resources.

Given the time frames required to achieve substantial water reform as pointed out earlier in this chapter to achieve the reform required in South Africa, it is too early to tell whether the outcomes of the reform process in Australia will all be successful.[11] However, what has happened is a good example of a relatively rapid adaptive policy and management response to water scarcity induced by over allocation and climate change. It is an experiment that should be closely watched by other countries in similar situations.

The Situation in Other Countries

As is to be expected, the recognition of water rights and the development of allocation, pricing, and trading policies in developing countries ranges from quite advanced in some Latin American countries including Mexico and Chile, to very limited in other jurisdictions.

Many countries are making good progress with the development of general water policies, legislation, regulation, and guidelines, but implementation and regulation of these is patchy, often as a result of poor understanding, corruption, and lack of policing.

In India ten years ago, only 10% of rural households were dependent on state-provided or community-managed organizations for water. The remaining households were either providing for themselves through privately owned and managed supplies of water or drawing from a government or community supply.[12] The dependence on other sources for water supply apart from the government is an indication of the range of systems operating to provide water and the importance of private sellers. For the poor in developing countries, more often than not it is these arrangements that are most relied on. In areas where formal institutional administrations have not been able to reach the poor, informal markets have sprung up to fill the gaps in services.

The Economics of Water

At the beginning of this chapter, we pointed out that microeconomic approaches to solving water scarcity and availability issues will not work in isolation. However, in environments where water access rights and allocation policies are well-defined, microeconomic approaches can be very helpful in determining which uses of water are likely to be most productive for the region and economy as a whole. Similarly, they may be extremely helpful in determining which water saving technologies are going to be most effective and worthy of investment.[13]

An irony of deepening water scarcity around the world is that in most developing countries, water is getting increasingly scarce but still remains free. One reason for this state of affairs is the informal nature of water economies in developing countries: As public or private service providers in the organized sector invest in water infrastructure and services, they begin to charge a service fee (though seldom a resource price) that approaches the cost of provisioning. Most small and large users still self-supply their water needs by directly diverting it from aquifers, rivers, and streams; these do incur a cost in water provision but are often far lower than the rising economic value of water in those societies. In most developing countries, it is also widely observed that the poorest segments remain outside the ambit of public water supply systems and end up depending on private water markets, paying rates far higher than what the rich pay or the economic value of water. In sum, a plethora of efficiency, equity, and sustainability issues in developing world water economies remain unresolved because of the roadblocks faced in aligning water prices with its economic value. Many international summits and bodies— such as the World Water Council —have recommended that water be treated as an economic *and* social good, yet progress in assigning economic value to increasingly scarce water remains patchy.

One reason perhaps is that there is no such thing a single "economic value of water" in a country, although we can find a small range of "cost recovery prices" (the cost of providing the service of water delivery) that are generally different from "economic value." The "true economic value of water" would vary across locations and for different purposes, even in a small country/state. In evolving a procedure for charging a "resource fee" for bulk water diverters, for example, the approach should be on identifying the "relevant cost" rather than historical or accounting cost, and the relevant cost will depend on the decision-on-hand and vary from decision to decision.

It is rare that water utilities employ economists in most developing countries;[14] as a result, water pricing is often informed by cost recovery that reflects historical cost rather than opportunity costs. An important need in most developing countries is to enhance awareness and generate debate about alternative concepts of "economic value of water." The following points need to be considered when looking at the real cost of water provision and how it might be priced:

- Economic full cost of water production (cost recovery price)
- Value of water in the next best alternative use in the location in question (opportunity cost)
- The loss of value/welfare by removing water from existing uses (agriculture, for example) to allow a new use
- The cost of replacing water diverted through enhanced water harvesting and recharge (local replacement or "cleaning up" cost)
- The cost of replacing water diverted through long-distance transport (basin level replacement cost)
- The cost of using an alternate source (such as desalinated water)
- The cost of mitigating external effects of water diversion

The real demand-side opportunity cost—by far the best indicator of economic value of water on the margin—may involve identifying the water price just high enough to drive away potential water-using investors; the price the government can charge can then be anywhere between the relevant cost-recovery price and this "what-the-market-can-bear" price. Unfortunately, even in developed countries (see Figure 9.1), many of these economic principles are ignored, and a "postage stamp" pricing approach is applied to water (i.e., the same price for delivery anywhere in the country irrespective of difficulty or distance).

Figure 9.1 Dethridge wheels have been used to measure water flow in irrigation schemes in Australia and the United States for approximately 70 years. They are used as the basis for determining on-farm water consumption and thus are the basis for water charging.

Photo: Colin Chartres

Marginal Cost Pricing Mechanisms for Water Allocation

If water prices are left ultimately to the market to decide, it is important that all costs are at least considered. These include the so-called externalities that occur in terms of the impacts of increased salinity in water after use, environmental damage, and so on. Marginal cost pricing mechanisms (MCPs) attempt to provide a means of dealing with these. There is a difference between traditional social/equity-based mechanisms for allocating water with approaches that treat it as an economic good and are based on economic efficiency. A number of important allocation criteria including flexibility, security of tenure, real opportunity cost, predictability of outcomes of the allocation process, equity and political and public acceptability, efficacy, and administrative feasibility and sustainability. From an economics perspective, marginal cost pricing defined as targeting a price for water based on the cost of supplying the last unit of that water, is considered an economically efficient, or socially optimal, allocation of water resources. Most water supply schemes are quota-oriented, and if prices are introduced, they intend to cover costs and are mainly based on average cost pricing. It is also important to consider that if there is a move to marginal cost pricing, there should be consideration of both social costs and externalities such as environmental degradation due to lower flows after extraction of water or higher salinities or temperatures when water is discharged back to the water body. While MCP is theoretically efficient and avoids underpricing and overuse of water, and thus is a potential way of dealing with water scarcity, it does have disadvantages. These include that it is multidimensional in nature and includes quantity and quality issues and that it can vary over periods in which it is measured and on whether a demand increment is temporary or permanent. It can also be argued that it tends to neglect equity issues and can be difficult to implement because it requires volumetric measurement and monitoring.[15]

Public Water Allocation Mechanisms

Most major irrigation schemes rely on publically funded water allocation mechanisms. This is because of the problems with people's perception that water is a public good and should not be treated as a commodity and that the levels of investment in large-scale water schemes are very high and returns low, thus deterring involvement of the private sector. Note also that when the private sector did take over some municipal supply schemes in cities including Buenos Aires and Manila, there was general outcry that water prices rose and infrastructure was not maintained. There is a strong case for public sector water allocation mechanisms, but there are also disadvantages. A particular concern is that subsidized supplies of water to water-scarce regions replace market mechanisms of water supply via transfer of titles (such as the example in the Murray-Darling Basin). This can lead to waste and overuse. Furthermore, government controlled irrigation systems are often poorly performing, and municipal systems operated by public utilities are often badly maintained and subject to major leakage losses. Similarly, licensing and regulation are often inadequate, which leads to water stealing and environmental contamination through inadequate management of discharges into receiving waters.

Market Mechanisms for Water Allocation

Water markets depend on clear legal identification of water rights, separation of water and land rights, the development of trading principles and rules, and the physical establishment of a market in which to trade—although there are many cases of more informal trading between individual users having occurred. Markets usually require many willing sellers and buyers all aiming to optimize profits.[16] When these conditions are met, they are effective ways to ensure that water moves from low-value to high-value uses. However, there are often social costs, particularly with respect to declining regional economic activity, in areas that are trading water out, and in some cases if several farmers on a branch canal sell their water elsewhere, this can make supplying a few remaining users extremely costly.

Paying for Pollution

Market-based instruments for pollution control are another way to value water and its quality. There are examples of polluter pay schemes from a number of countries around the world. A typical example is that of salinity in the Murray-Darling Basin. Here a benchmark salinity level is used for the Murray River at Morgan (the offtake point for Adelaide's water supply). Salinity concentrations and loads and actions that will affect them above this point in the main streams and their tributaries are then assessed against end of valley targets. An action will be considered as significant and included in the Commission (at that time the MDBC) Registers if it is assessed to cause a change in average EC at Morgan of 0.1 EC or higher within 30 years. Salinity credits and debits are the mechanism by which end of valley targets can be met based on the principle that increasing salinity at one point has to be offset by reductions elsewhere. For example, increases (debits) might arise from saline discharge from industry or agriculture, while decreases (credits) can be gained by improved land management practices and/or interception of saline aquifers discharging into the river.

Lessons from Complexity?

For many in the water business, definition of water rights, allocation policies, and water pricing mechanisms are fraught with complexity and difficulty. They also lead to considerable public concern and mistrust of governments, and as a consequence, water policy is often an area that reform has passed by, particularly in comparison with public utilities like railways, gas, and electricity. However, increasing water scarcity means that this cannot be the case in the future and emphasizes the pressing need for reform of water governance and policies in many countries.

Some basic principles clearly arise from the material presented in this chapter. Whatever route countries choose to deal with water scarcity depends on their dealing with these principles. They are

- Water rights have to be legally defined.
- This includes rights of access to water for all and enshrinement in legislation of the right to sufficient water for drinking, cooking, and washing.

- The environment should also have its rights to water defined and protected. If we want to protect ecosystems and valuable ecosystems services, it is not good enough to just allocate to the environment what water is left after all other users are satisfied.
- Ideally water rights should be separated from land rights to enable redistribution and trading of water.
- Water should be valued appropriately depending on its scarcity.
- The real cost of water provision should be calculated and considered in terms of users' ability to pay.
- Over and above the provision of water for basic human needs, commercial users should be charged for water at a level commensurate with its cost of provision.
- What can't be measured can't be managed. Water flows and water use should be adequately measured and metered to facilitate best-practice management.
- In water-scarce environments, allocation mechanisms such as marginal cost pricing should be considered as ways in which water use is optimized and overuse reduced.

Many of these principles are already partially enshrined in country legislation, but as has been pointed out, regulation and policing is not being carried out effectively. Our view is that water is totally "undervalued" in that its previous ready availability made many of us take it for granted. There needs to be a paradigm shift that has us begin to see water like we see oil. We are close to the point in time of "peak oil," after which production starts to decline, and yet as we all know demand continues to increase. The automotive industry and governments are investing billions of dollars in solutions to this issue, spurred on by the added threat burning fossil fuels has on climate change. It is highly likely that if we had a concept of "peak water," we have passed this point in a number of countries already. Yet investment in R&D and innovation in the sector that uses the most water, agriculture, has declined significantly since the 1980s. Clearly, the messages coming from science are not being heard loud enough or acted on sufficiently in most countries. One reason for this is that those countries worst affected are not in the developed world, with the possible exception of Australia and the Southwestern United States. However, not having enough water to feed growing developing and emerging nations will potentially wreak more havoc than not having enough oil—havoc that will spread around the globe in terms of social and political unrest, mass migration, and potentially terrorism. It is time developing country governments woke up to this reality. However, it is not only developing country governments that need to wake up to these stark realities—we all do. One way to do this will be to examine what are the types of incentives that will bring about the changes in water policy and governance that will affect required changes in usage. The nature of these incentives are discussed in the concluding chapter.

Conclusion

This chapter exemplified two key points. The first is that developing effective water rights legislation and pricing mechanisms is something that takes decades if not centuries to achieve. The second point is that these systems are extremely complex, and laws vary considerably from place to place even within countries. Our concern is that given the magnitude of the water crisis and the accelerating effects of climate change, many developing countries will not have the time, money, or patience to allow evolution of their water rights and markets legislation before it is too late. This means that we should always be on the lookout for "leapfrogging" solutions. A good analogy here is the way in which access to mobile phones transformed communications in

Africa, where landlines were poor and had limited reach. We are not advocating that everybody should pay for water but that it is important for all to realize that water does not come for free and that there are some principles to follow on which improved provision of water rights and pricing policies can be based.

THE KEY OBJECTIVES OF THE AUSTRALIAN WATER ACT 2007

a. To enable the Commonwealth, in conjunction with the Basin States, to manage the Basin water resources in the national interest; and

b. To give effect to relevant international agreements (to the extent to which those agreements are relevant to the use and management of the Basin water resources) and, in particular, to provide for special measures, in accordance with those agreements, to address the threats to the Basin water resources; and

c. In giving effect to those agreements, to promote the use and management of the Basin water resources in a way that optimises economic, social and environmental outcomes; and

d. Without limiting paragraph (b) or (c):

   i. To ensure the return to environmentally sustainable levels of extraction for water resources that are over-allocated or overused; and

   ii. To protect, restore and provide for the ecological values and ecosystem services of the Murray-Darling Basin (taking into account, in particular, the impact that the taking of water has on the watercourses, lakes, wetlands, ground water and water-dependent ecosystems that are part of the Basin water resources and on associated biodiversity); and

iii. Subject to subparagraphs (i) and (ii)—to maximize the net economic returns to the Australian community from the use and management of the Basin water resources; and

e. To improve water security for all uses of Basin water resources; and

f. To ensure that the management of the Basin water resources takes into account the broader management of natural resources in the Murray-Darling Basin; and

g. To achieve efficient and cost effective water management and administrative practices in relation to Basin water resources; and

h. To provide for the collection, collation, analysis and dissemination of information about:

i. Australia's water resources; and
ii. The use and management of water in Australia.

AUSTRALIA'S NATIONAL WATER COMMISSION VISION FOR A BETTER WATER FUTURE (NWC 2006)[17]

Shared National Objectives

1. The National Water Initiative continues in place as the agreed national blueprint for water reform in Australia.
2. Cost-effective water use efficiency is adopted throughout Australia as a unifying national objective in water use and management.

Highly Effective Water Planning and Decision Making

3. Water planning processes throughout Australia are fully participatory; consultation with stakeholders is thorough.

4. Production, environmental, and social objectives are all taken into account.
5. Best available science and economics are used for planning and decision making.
6. The necessary investments are made in acquiring scientific and economic data, knowledge, and information sufficient for sound decision making. Water data are openly shared among all parties.
7. Australian water users, including environmental managers, enjoy certainty and clarity about their water entitlements.
8. Australia's approaches to water regulation and management are nationally harmonized among the states and territories at best practice level.

Environmental Sustainability

9. The need for sustainable use of Australia's water resources is taken as given, and judgments about environmental sustainability of Australia's water systems are made in an integrated fashion with decisions about entitlements and allocations for production purposes.
10. Water management is integrated with wider natural resource management.

World Class Water Management

11. Best available water management techniques and technologies are actively sought and implemented.
12. Throughout Australia adaptive water management practice is employed to reflect what has been learned.

13. Australia's irrigation industry is seen as a model for the rest of the world in terms of water use efficiency and sustainability; irrigation technologies and practices are tuned to Australia's unique circumstances.
14. Australia is recognized as a world leader in water reuse and recycling.
15. Water markets and water trading are widespread throughout Australia and considered "normal."
16. There are expanding opportunities for private sector involvement in the Australian water industry.

chapter ten

Solving the World's Water Problems

"Thousands have lived without love, not one without water."

W. H. Auden

We have a looming water crisis. This crisis is the response to growing population, changing dietary habits, and competition for water from other sectors of the economy. Lack of water for growing food will be one of the most critical issues for us to overcome in the twenty-first century. Evidence that the water crisis is happening in some places already or that it will happen elsewhere comes from simple supply and demand comparisons and projections. We are now faced with the knowledge that climate change will accelerate the water crisis in many regions of the world and have a disproportionately serious effect on the poor. Many knowledgeable scientists still argue that we don't really have a crisis because current water scarcity is caused by inefficient water use. To an extent this is true and provides a basis for hope in that if we can improve efficiency, we can deal with many of the problems. However, the fact is that while we know this, trends in water use efficiency and water productivity in agriculture, the major consumer, are not very encouraging particularly in the developing world. Furthermore, many biophysical and engineering solutions to increasing productivity are also available. It is not as if we are waiting for the next scientific breakthrough to occur. So we have to look more closely at socio-economic factors to determine what might be done to get us on the water use efficiency bandwagon. Scientifically, this is easier than dealing with the energy crisis, but in terms of applicability, solutions are severely handicapped by people just not understanding the potential severity and magnitude of the impending crisis and as yet, generally not altering the ways in which they view and use water. If we don't succeed in changing the way we manage water, the future looks bleak and will consist of increasingly frequent food crises, social and political unrest, and potential mass migration out of areas most severely affected.

We need to consider an action agenda that will promote a so-called "Blue Revolution." We perhaps ought to call this a Blue-Green Revolution because it is imperative that we do not ignore the use of green water for rainfed agriculture. To an extent, there are examples of regions and projects that have been successful in implementing water reform, but the task ahead of us is to ensure that these successes are repeated time and time again across the globe. To do this we need a change in the way people think about and view water. It has to be viewed and thought about on a similar basis as we currently consider energy supplies.

In the following sections, we lay out a blueprint for water reform. It draws on several of the previous suggestions detailed in the Comprehensive Assessment of Water Management in Agriculture[1] and adds additional thinking and material. This blueprint is focused around six critical challenges:

1. If you can't measure it, you can't manage it.
2. Treasure the environment.
3. Reform water governance.
4. Revitalize agricultural water use.
5. Manage urban and industrial demand.
6. Empower the poor and women in water management.

1. If You Can't Measure It, You Can't Manage It

This simple yet essentially true saying holds applies to water, as it does for most goods and services. Indeed, one of the reasons that so many countries have got into a mess with their water management is that there is not enough attention paid to determining the flows and storages that make up their water resources. Furthermore, as a wave of divestment of so-called noncore activities hit the western economies in the Reagan-Thatcher era, many governments sold their water utilities and monitoring networks to the private sector. Not surprisingly, when the private sector looked at the state of some of the infrastructure and the investment required to maintain it, they abandoned or reduced the scope of some previous activities. One of these was often measurement and monitoring of the resource. At the same time, previously publicly available data were sometimes made inaccessible because it was now considered to be commercially-sensitive and thus valuable. So by the turn of the twenty-first century, many developed countries entered the new water era with less than adequate information about their water resources. Developing nations have never had the opportunity and resources to invest in monitoring networks to the extent required, and as a consequence there is often a dearth of useful data. The situation is also made more difficult in some countries by an unwillingness of governments to share water data with the public and science agencies on the basis that it is of high national security value and must be classified. This is essentially nonsense, as the basis of establishing transboundary water agreements has to be common understanding of what the resource consists of in the first place.

Access to high-quality data and information should be the goal of all water planning and management departments. These data can provide essential information about responses to rainfall events, flood hazards, seasonal flow variations, groundwater level changes, and the impacts of major extractions on flow. Ironically, as hydropower generation becomes increasingly important, those who produce hydropower are vitally interested in water availability and discharge data, particularly where they may be competing in the market place for electricity supply. Their measurements of water flow are often recorded every few minutes. When these data are accumulated to daily or weekly totals, it has less commercial interest and should be made readily available to downstream planners and managers.

In 2006, the Australian National Water Commission indicated in a press release[2] that open access to water data will

- Ultimately reduce costs to data users, including transaction costs
- Reduce data inconsistencies, data gaps, and lack of comparable data
- Enable performance benchmarking
- Enable national water assessments on a repeatable basis
- Enable better water planning, including cross-border
- Underpin markets
- Redress declining community confidence in the national water market
- Reduce multiple requests for information to data custodians
- Reconcile the sometimes conflicting data needs of water data gatherers, managers, and users

When this list is analyzed, it becomes apparent that virtually all the points can help to lead to greater efficiency of water use. It goes without saying that similarly having appropriate data is the best way in which to discuss transboundary water sharing, provide a basis for water allocations, and also to provide a moving picture of the impact of climate change and additional water use on the water resource and the environment as a whole.

While water gauging and monitoring can be a time-consuming and expensive exercise, developments in technology are making things cheaper and easier all the time. Once installed, many gauging stations and groundwater bores can be monitored remotely by satellite or mobile phone linkages. Similarly, database and geographic information technologies have made the storage, retrieval, and spatial display of information straightforward, and most information can be portrayed at the press of a button or mouse-key. Exciting new advances in remote sensing technologies have recently opened up the possibility of monitoring major changes in groundwater levels using gravity data.[3]

All currently water-scarce countries and those approaching scarcity need to consider how they can develop water measurement and monitoring systems that will underpin water management at national, regional, and local levels. The degree of investment and sophistication will depend upon the ultimate needs of water managers and users. There is one final word of caution here. That is that under climate change scenarios, the risk of relying on past water data may be an inappropriate way to plan for the future. Scientists and engineers cannot rely on what is called the *principle of stationarity* and will have to build improved downscaled climate change predictions into their water models to cope with this issue.

2. Treasure the Environment

It is strikingly clear that a growing water quality problem accompanies water scarcity. While many western nations have made very significant inroads in cleaning up rivers highly polluted by industrial effluents and human sewage, this is still not the case in many developing countries. Furthermore, pollution issues now also include the threats posed by pesticides and persistent organic pollutants (POPs). The UN Environment Program describes the latter as

> "...chemical substances that persist in the environment, bioaccumulate through the food web, and pose a risk of causing adverse effects to human health and the environment. With the evidence of long-range transport of these substances to regions where they have never been used or produced and the consequent threats they pose to the environment of the whole globe, the international community has now, at several occasions called for urgent global actions to reduce and eliminate releases of these chemicals."

In Sri Lanka, for example, overuse of fertilizers, encouraged by high government subsidies, and the use of pesticides cause pollution of water courses and ultimately coastal lagoons. Eutrophication caused by the fertilizers can lead to algal blooms and their choking impact on aquatic life, whereas the POPs can impact all life forms in the food chains dependent upon the waterways affected and also subsequently marine life.

It is important that we start to think about the environment not just as the receiver of human effluents, but as a highly valuable series of ecosystems that if treated properly can play a very significant role in the cleaning and decontamination of wastes and in providing sustainable supplies of fresh uncontaminated water. A well-known Australian hydroecologist, the late Professor Peter Cullen, described the first mile of river downstream of a sewage treatment plant as "the magic mile," magic because sewage was treated by dilution and biological processes to such an extent that people were prepared to extract the water further downstream for drinking purposes. A common saying is "dilution is the solution to pollution." While dilution processes may hold for small discharges and large rivers, unfortunately the quantities of sewage are now often so great that they totally befoul rivers and destroy their wildlife. In the worst cases, virtually all living materials are being destroyed, and all that is left is a black, stinking, anoxic flow of water.

However, we know that this does not have to be the case. Regulations and economic incentives and fines can be used to limit point source discharges, and improved agricultural management practices including minimizing runoff and erosion can be used to reduce non-point pollution. However, these often do not work as well as they should because of a combination of factors. Flouting the law in developing countries is commonplace, and often the authorities do not have the human resources or authority backed up by the judiciary to ensure compliance. Corruption also often raises its ugly head in terms of allowing polluting industries to establish themselves in totally inappropriate areas.

So two key issues have to be dealt with: The first is that non-biodegradable and endocrine disrupting substances (including POPs, heavy metals, and certain other chemicals) should never be allowed to enter the hydrological cycle unless they are either proven to be safe or diluted to such levels that any risks are minimal. The second is that rivers need to have a portion of flow set aside for the environment. These flows should be managed in such a way that they emulate natural conditions in terms of peak and low flows and overbank discharges that flood the floodplain. Environmental flows can also be managed in such a way that water used for the environment in one section of the river can often be used for other purposes downstream.

3. Reform Water Governance

Technological and engineering solutions to double food and feed production are the easier part of the equation to solve. Overcoming the social, economic, and sometimes environmental impediments and obtaining the needed financial investment is the hard part. Institutional and governance arrangements often were designed in the middle of the last century based on inappropriate colonial models where water was viewed as an infinite resource.

Even if they are renewable, water resources are finite. A new governance paradigm is needed to meet the challenge of feeding growing populations (see Table 10.1).

If the challenge to feed more with less was not great enough, the shift from planned and regulated (albeit inadequately) surface irrigation systems to anarchic pump-based irrigation systems based predominantly on groundwater that has occurred in South and East Asia threaten to literally dig us into a deeper water hole. The inability of governments to regulate water use in such systems can create the scary scenarios of groundwater overdraft and exhaustion. These can in turn lead to regional food crises and social disruption.

Table 10.1 *Changes in Water Resource Governance Expected as Basins Move from Open to Closed**

Open Basins	Closed Basins
Exploiting water resources	Managing demand
New allocations	Reallocating water
Who is included and excluded	Safeguarding right to water
Developing groundwater	Regulating groundwater
Institutions for single sectors	Institutional frameworks able to deal with cross sectoral issues
Within system conflicts	Cross sectoral conflicts

*With thanks to David Molden.

Governments lack incentives to implement the reforms necessary to ensure more productive and equitable use of water. Fear of potential political repercussions for those who push reform permeate the water and agricultural sectors from top to bottom.

To develop incentives and support for reform, water has to be seen as something that can be valued—and ultimately priced. It cannot continue to be treated as a "free" good. This does not mean that the human right to water is overlooked in the process. Few would argue against access to clean water for drinking and sanitation being a fundamental human right that must be protected in any wholesale change to the way water is governed and managed. However, this human right accounts for a very modest amount of total water use. The rest, probably about 90%, goes to beneficial uses and the environment. The biggest beneficiary is clearly agriculture.

Measures that governments can take to drive up agricultural water productivity are nonexistent in many countries. Clearly the first measure has to be the development of effective water allocation policies, which can be used to reduce allocation as the total pool shrinks or when demands for water resources from other sectors increase. However, allocation policies depend on good water availability measurements, historical data, and models as well as defined water rights. Reduced allocations must be accompanied by support mechanisms for farmers that can improve on-farm efficiency. Currently, if a farmer invests in improving productivity, he or she can keep the water saved and use it to increase the area irrigated. While this may increase food production, it does not solve the problem of reallocation of water to other economic sectors or to the environment. A real challenge here is to try and develop incentives that link the broader society to farmers and lead to the broader society paying farmers for the improved environmental services and other benefits that result from improved on-farm water savings.

In the search for improved governance, we must examine the potential solutions that have been and are currently being developed. In parts of Australia and several other countries, a series of mechanisms are used to regulate water use and allocation that depend on seasonal available supply. In the Murray-Darling Basin of Australia, a new system of separation of water and land rights, water trading, and water pricing based on supply and demand, albeit under an overall planning framework, has evolved through a combination of market and political forces. The result: water is traded from low- to high-value uses, which can potentially allow for a market mechanism for trade out of water from agriculture into urban areas. It is a model worth exploring elsewhere. So long as individual water rights and allocations can be defined, it provides farmers opportunities and incentives to sell temporarily or permanently. It also gives governments opportunities to buy out system tail-end users (those irrigators at the end of the distribution system), improve overall system efficiency, and to buy water for environmental flow purposes.

Water scarcity is an increasingly urgent challenge to the developing world. It can be combated, but it needs politicians and policy makers to develop some enthusiasm for reforming the water sector. Developing appropriate market-based and other incentives is vital to reform in the water sector. Better definition of water rights and better measurement of water are needed to even contemplate better systems for valuation, pricing, and trade. Without these improvements, there will be few incentives to improve productivity whether by the use of economic or regulatory instruments.

4. Revitalize Agricultural Water Use

An aspirational goal for agriculture everywhere is to produce twice the yield of half the area and use no more water than at present. If this goal can be achieved and the productivity of developing country water and agricultural systems doubled, it is likely that food security issues would be a thing of the past as long as population stabilizes. In many respects, doubling crop yields should be easy in many countries. For example, the productivity of water in Pakistan is among the lowest in the world as exemplified by crop yields that are up to three to four times lower than those in developed countries like the United States. This reveals a substantial potential for increasing the productivity of water. However, lifting productivity and increasing efficiency are often extremely complex tasks in all countries. Generally, while the same factors including water availability, quality, pumping costs, soil properties, agronomy, fertilizers, pests and diseases, and farmer capacity are the same controlling factors everywhere, the way they combine differs substantially from place to place. Thus, low productivity in one area might be due to soil conditions, whereas across the fence it might be limited by lack of fertilizer, inadequate water management, and/or other factors.

However, there is good evidence from many countries that with improved knowledge sharing, improved management practices, and appropriate market incentives, productivity can be raised significantly. Research and development undoubtedly plays a major role in helping lift yields and water productivity. However, as demonstrated in an earlier chapter, investment in R&D in agriculture as a proportion of total overseas development aid has decreased significantly since the 1980s and may be a principal reason for the reduction in the rate of yield increases particularly in developing countries, which are not pushing against biological potential. This decrease in investment was a response at the time to the success of the green revolution. However, what aid agencies forgot was that the creeping tide of population growth was swallowing up the yield increases stimulated by the green revolution at a rate faster than yield growth.

If we look specifically at improving agricultural water management as part of the broader equation of lifting agricultural yields and productivity, a number of solutions are already apparent. These can be viewed across national, regional, and local scales. At the global regional and national levels, a recent report released by IWMI and FAO[4] has highlighted the significance of irrigation to food production in Asia where 34% of the cultivated land is irrigated in contrast to 10% in North America and 6% in Africa. This essentially was a major reason why the green revolution was so successful in parts of Asia. Irrigation facilitated a 137% increase in cereal production between 1970 and 2007. However, the report points out that many of the Asian irrigation schemes were built in the 1960s and 1970s and are aging and in need of significant rehabilitation. Also the face of Asia is changing fast, with more urbanization, greater opportunities for people, and in irrigation a shift from surface water systems to groundwater (as described in Chapter 5, "Agriculture and Water"). A key recommendation not only of this report, but one we believe is imperative is that governments in Asia must review the extent, status of their irrigation systems including their productivity if food production targets are to be met. They must also give attention to how they can harness and appropriately regulate, with respect to sustainability, the growing extraction of groundwater, and they must ensure that the institutions managing their irrigation systems focus on providing farmers with the services they want and when they want them, rather than bureaucratic convention.

In Africa, the opportunity is to invest in new irrigation schemes. There has been a reticence to do this because of past failures. Growing food demand will, however, provide the incentives that farmers need to drive a new wave of irrigation. However, it does not have to be based on the old twentieth-century models that have been seen to have major shortcomings in Asia and elsewhere. Irrigation does not have to be implemented on massive scales with a monocultural (year-in, year-out cropping with one crop type) cropping focus. It can work and improve livelihoods and production at smallholder level with supplies being drawn from rainwater harvesting, groundwater, or small ponds and reservoirs with a focus on supplementary irrigation of both staple crops and cash crops such as vegetables to diversify diets and insure against crop failure. There are already many examples of, often,

NGO (non-governmental organization) driven projects that focus on multiple use water systems that have succeeded in providing both supplies of domestic water and some limited water for irrigation of small plots. These kinds of developments, if adequately supported in terms of development of the entire value chain from water provision to marketing of produce and backed up with capacity building with respect to parts supply and maintenance, are beginning to show major success in Africa and Asia.

In some respects, increasing water productivity in rainfed agricultural areas is a more difficult challenge. These areas are subject to the increasing variability of rainfall as well as increasing temperatures caused by climate change. They also often suffer from low fertility and land degradation issues. Consequently, smallholders are often impoverished and live on their crops and animals in a hand-to-mouth manner. Looking at whether improving water supplies and the provision of opportunities for supplementary irrigation is one solution for some of these areas. Other solutions include looking at how soil water can be conserved and protected from runoff and evaporation losses. Given the areal extent of rainfed agriculture, particularly in Africa, there have to be significant productivity increases from these areas. We cannot simply rely on increasing productivity of existing irrigation areas in Asia and building new irrigation systems in Africa. Furthermore, dealing with rainfed productivity issues also brings a significant opportunity to increase the livelihoods of some of the world's poorest people.

Recycling of wastewater will also have an increasingly important role to play in increasing water productivity. However, the issues regarding the development of safe practices have to be given significant attention. Studies point out that policies and decisions on wastewater use in agriculture should generally be motivated locally, as the socio-economic, health, and environmental conditions, which vary across countries, will dictate how far common recommendations are applicable.[5] They also developed some rules and guidelines for wastewater use that are shown in the following sidebar.

RECOMMENDATIONS AND GUIDELINES FOR SAFER USE OF WASTEWATER IN PERIURBAN AND URBAN IRRIGATED AGRICULTURE[6]

1. **Deal with knowledge gaps.** The gaps in knowledge of the true extents of the often informal use of wastewater at a country level must be addressed by governments through detailed assessments, which will allow them to evaluate trade-offs and decide on the hot spots that need immediate attention.
2. **Ensure World Health Organization (WHO) Guidelines are applied.** The WHO guidelines for the safe use of wastewater[7] should be extensively applied as it allows for incremental and adaptive risk reduction in contrast to strict water quality thresholds. This is a cost-effective and realistic approach for reducing health and environmental risks in low income countries.
3. **Link water supply and sanitation sectors.** Implementation of the Millennium Development Goals should more closely link policies and investments for improvements in the water supply sector with those in the sanitation and waste disposal sector to achieve maximum impact.
4. **Separate industrial and domestic discharges.** To improve the safety of irrigation water sources used for agriculture and enhance the direct use of wastewater, it is imperative to separate domestic and industrial discharges in cities and improve the sewage and septage (partially treated waste) disposal methods by moving away from ineffective conventional systems.

5. **Understand health risks.** A research gap clearly exists on quantitative risk assessment studies, which include multiple sources of risk, and such studies must be commissioned at a city or country level before decisions are made on water and sanitation sector investments.
6. **Apply simple hygiene rules.** Acknowledging that off-farm handling practices like washing of vegetables can be very effective as a means of reducing/eliminating contamination, and supporting widespread use of good practices can facilitate trade exchanges for developing countries exporting vegetables.
7. **Develop economic and financial incentives.** In addressing health risks, on the one hand state authorities have a role to play in planning, financing, and maintaining sanitation and waste disposal infrastructure that is commensurate with their capacities and which responds to agricultural reuse requirements. On the other, as a comprehensive treatment will remain unlikely in the near future, outsourcing water quality improvements and health risk reduction to the user level and supporting such initiatives through farm tenure security, economic incentives like easy access to credit for safer farming and social marketing for improving farmer knowledge and responsibility, can lead to reducing public health risks more effectively while maintaining the benefits of urban and periurban (partially developed areas around fringes of towns and cities) agriculture.
8. **Develop local policies and regulations.** Finally, countries must address the need to develop policies and locally viable practices for safer wastewater use to maintain its benefits for food supply and livelihoods while reducing health and environmental risks.

5. Manage Urban and Industrial Demand

A great deal of progress has been made already in this area, particularly in water-scarce countries. Water utilities have been able to convince and to some extent coerce their customers through media campaigns about water shortages and pricing water according to consumption. The history of what happened in Sydney, Australia, in this regard is a good example. Sydney's major water storages were built in the 1950s and early 1960s, and they were designed from rainfall data from a period that in retrospect was wetter than the present. Australia has always been subject to considerable climate variability, and the dry periods in southern Australia now appear to be being amplified by climate change impacts. Sydney has put in place a number of demand management measures over the last few decades. These measures included a series of steps that include replacement of leaking infrastructure, restriction of times at which gardens can be watered, banning the use of automatic sprinklers, prohibiting car washing except at commercial premises that recycle the water, and so on. During 2008–2009 Sydney has had better rainfall, which has allowed more severe restrictions to be lifted and replaced with Water Rise Rules (see the following sidebar). Additionally, there have been campaigns to reduce water usage in the home by fitting dual flush toilets and water-save shower heads, not to mention providing information about how much water is wasted by long showers and brushing one's teeth with the tap running. Complementary to these physical interventions have been increases in water prices. In comparison with inflation these have been quite small, and there is concern that pricing policies can often hit the poorest hardest and have little impact on water use by the more well-off sections of society. Overall, these measures have been combined with targets to recycle water to supply 12% of the city's needs by 2015. This water will only be used for industry, irrigation of parks, gardens, and golf courses and to provide environmental flows in the major river systems. As population in Sydney grew, supply was able to keep pace with demand largely because of a series of demand management practices put in place by the city water authorities. However, because of the still increasing city population and climate change, Sydney has also had to commit significant investment to the development of a seawater desalination plant, that by 2015 will be able to provide 250 million liters (66 million gallons) of water per day (15% of demand), using renewable energy as a power source.

SYDNEY WATER CORPORATION WATER WISE RULES

- All hoses must have a trigger nozzle.
- Hand held hoses, sprinklers, and watering systems may be used only before 10 a.m. and after 4 p.m. on any day to avoid the heat of the day.
- No hosing of hard surfaces such as paths and driveways is allowed. Washing vehicles is allowed.
- Fire hoses may be used for fire fighting activities only.

In other Australian cities and elsewhere, new movements focusing on water-wise urban design principles have been developed that aim to minimize losses to stormwater and promote recycling of urban runoff. In California, Orange County has lead the way with recycling sewage and injecting the treated water into the aquifer. Initially this was in order to minimize saltwater intrusion from the ocean, but now apparently, the aquifer water is being used for public water supply. This form of reuse is described as *indirect potable reuse*. It often causes considerable public controversy and is often opposed by sections of the community ranging from firefighters, to plumbers, to occasionally the medical profession on the grounds that we can't be sure that all contaminants and pathogens are removed. In one community,

Windhoek, the capital of Namibia, which has severe water scarcity, water recycling has been taken a step further.[8] There, treated sewage is recycled directly back into the water purification plant, where a combination of activated carbon filtration, ozonation, and pressure membrane filtration supplies potable water back into the urban pipe network. Here it is mixed with reservoir or groundwater. This system of direct potable reuse has been operating since the late 1970s and had a major upgrade about 10 years ago. There have been no incidences of illness associated with drinking water in the city in this entire period.

Most industries that use significant quantities of water have been aware of their need to monitor and reduce water consumption for several years, and some major progress has been made in this regard. Their interest in minimizing water use comes from an understanding of water scarcity, concerns that a sustainable water supply increases their own sustainability, and in some cases in response to corporate social responsibility objectives. In many cases, such as in the mining industry, a viable supply of fresh water may be a critical input into processing, and similarly, release of treated clean water into the environment may be fundamental to their ongoing survival under increasingly tough environmental legislation in more advanced countries. So the trend toward water awareness and water saving in industry has been one that initially looked at the amount of water used in their factories and processing plants with a view to minimizing the liters of water used per liter or kilogram of product. More recently, industry has been looking at the concept of water footprinting in terms of the total water demand not only in the product processing, but in its growth (if a foodstuff) and subsequent preparation for use/consumption. This work is in its infancy but promises to assist in highlighting inefficient water use and thus the introduction of best practice measures to deal with it. These programs are being actively supported by agencies such as the World Economic Forum and the International Finance

Corporation of the World Bank Group. In some cases industry leaders are becoming champions of efficient water use and speaking out about the need for it at a range of public venues and high-level fora.

It would be fair to say that the concepts of demand management, water reuse, and water treatment described here are predominantly those of developed and intermediate economies. However, they provide lessons for the way forward for cities and industries in the rest of the world.

6. Empower the Poor and Women in Water Management

Water is crucial to the lives and livelihoods of farmers across the world. The conditions under which water and poverty intersect are complex and deeply intertwined. Water is not just the key to improving the productivity of agriculture for the poor who own small shares of land and live on subsistence agriculture; it is also crucial at the larger scale of achieving food security to prevent the deaths of millions from starvation and malnourishment.

The question of equity in how resources are shared between people belonging to different groups such as women, children, particular minorities, or historically vulnerable groups is something that is being taken more seriously in water management. The representation and voice of the poor and marginalized in water infrastructure development and planning has a poor record as past efforts to build and invest in them was focused more heavily on construction and less so on the impacts on people or the environment. Although irrigation has made significant impact on poverty alleviation, by improving the productivity of vast regions particularly in Asia, smallholder farmers that depend largely on rainfed farming for their livelihoods need a greater investment through programs and interventions to improve their productivity. Farming in vulnerable environments such as uplands that are most susceptible to shocks and stresses caused by climate change and environmental degradation is usually performed on a small scale and dominated by a large number of the rural poor. These regions have the potential to be far more productive, and efforts to improve these need not be high-cost. Simple technologies to improve productivity and also to store water can make a very large difference to the lives and livelihoods of the poor in these regions, while also contributing to improving the food security challenge.

The question of rights and access to resources for the poor is also critical to the discussion on water management. Although in many countries this is a deeply political issue that often brings in the problems that power and politics play in aiding or pushing out vulnerable or marginalized groups of people, there is a strong need for institutions to play a better role in facilitating dialogue to address equity. Efforts to do this within institutions such as river basin organizations, water treaties, and in the assessments of the impacts of water infrastructure are increasingly being given more importance in water development. However, these concerns are only as good as the political will behind them, and as debate about the cost and valuing of water comes into the fore, particularly in water-scarce areas, so must the debate about securing water for the poor.

The role of women in water management is most often described in terms of the amount of time women spend collecting and carrying water. While this is an important aspect of the critical role of women, the productive contributions of women's labor in agriculture and food production is one that needs to be paid more attention. Many efforts at poverty alleviation involving technology have not adequately paid attention to women farmers by excluding them from decision making. Addressing the fact that women are involved in food production and play a large role in agriculture and as users of water is imperative to addressing poverty as well as food security (see Figure 10.1). Policies and investment need to be more proactive in addressing the inequities of resource allocation. Without a direct effort to improve the distribution of water and other inputs into agriculture, there will not be a serious impact on poverty. Many development organizations are now targeting their programs to the poor, women, and smallholders in an attempt to address equity. The approach of governments to addressing poverty among farmers has largely been based through welfare programs such as food-for-work on water infrastructure, and with the idea to provide safety nets. While many of these have been successful, it is important to remember that agriculture can be made more profitable through relatively small but targeted investments.

Figure 10.1 In many countries, women do the bulk of agricultural work, but societal "norms" often mean that their voice is not heard. Recognizing the key role of women in agriculture is a fundamental requirement for water governance. Here, women farmers discuss key issues in Gujurat, India.

Photo: Sharni Jayawardane

Conclusion

In this chapter, we distilled a large amount of material and concepts into six actions that will help address the imminent water crisis. Key challenges and actions are shown in Table 10.2.

As we have shown, agriculture is in most countries the major user of water. In a world where increasing competition for water and negative climate change impacts are going to reduce the water available to agriculture in many countries, it seems logical that most attention then has to be paid to making agriculture water use more efficient and productive. However, as we have shown, while this is not rocket science, it will only be achieved through an integrated approach that combines the best science and engineering with first class economic and social policies that create the reforms needed to empower all in the water sector. In the past societies and science have never dealt brilliantly with similar complex issues. Indeed, science has thrived on reductionist principles that provide explanation to smaller and smaller parts of our natural environment. If we are to solve the closely related problems of water scarcity and food security, we have to move forward with a truly integrative approach to these mighty challenges. The problem is that we have a very limited time frame in which to do this.

Table 10.2 *Actions Required to Underpin a Blue Revolution*

Critical Challenge	Required Actions
1. Water information ("If you can't measure it, you can't manage it.")	▪ Establish water monitoring networks at river basin/country level
	▪ Make water data freely available to all
	▪ Plan on the basis of evidence
	▪ Take potential climate change into account

Table 10.2 *Actions Required to Underpin a Blue Revolution*

Critical Challenge	Required Actions
2. Treasure the environment.	▪ Ensure all water systems have water allocated to environmental flows ▪ Regulate and police pollution ▪ Implement paid ecosystem services
3. Reform water governance.	▪ Improve identification of water rights and allocation systems ▪ Govern and manage water on the basis of Integrated Water Resources Management principles ▪ Value water appropriately and consider market-driven opportunities for reform
4. Revitalize agricultural water use.	▪ Increase water productivity in irrigation ▪ Improve soil water holding capacity in rainfed areas ▪ Consider agricultural systems on a farm-to-fork value chain basis to improve profitability
5. Manage urban and industrial demand.	▪ Introduce new technologies ▪ Use suite of demand management strategies including regulations and pricing ▪ Increase recycling and reuse
6. Empower the poor and women in water management.	▪ Provide positive incentives to break the cycle of water poverty ▪ Make institutions more inclusive

Endnotes

Chapter 1

[1] J. H. Hamner and A. T. Wolf, "Patterns in International Water Resource Treaties: The Transboundary Freshwater Dispute Database," *Colorado Journal of International Environmental Law and Policy*, 1997 Yearbook (1998).

[2] Gethin Chamberlain, "India prays for rain as water wars break out," *The Observer* (July 12, 2009), http://www.guardian.co.uk/world/2009/jul/12/india-water-supply-bhopal.

[3] Comprehensive Assessment of Water Management in Agriculture, *Water for Food, Water for Life. A Comprehensive Assessment of Water Management in Agriculture*. D. Molden (ed.), (London: Earthscan and Colombo: International Water Management Institute, 2007).

Chapter 2

[1] F. Molle, P. P. Mollinga, and Meinzen-Dick, "Water, Politics, and Development: Introducing Water Alternatives," *Water Alternatives* 1(1) (2008):1–6.

[2] This saying adorns the foyer of the International Water Management Institute in Colombo. While it echoes important sentiments about appropriately valuing water, we are never sure of its political correctness given the fact that it does not seem to take the need for environmental water into account.

[3] R. Courcier, J.-P. Venot, and F. Molle, "Historical Transformations of the Lower Jordan River Basin (in Jordan): Changes in Water Use and Projections (1950–2025)," *Comprehensive Assessment Research Report* 9, International Water Management Institute, Colombo, Sri Lanka (2005).

[4] Ibid.

[5] T. Shah, *Taming the Anarchy: Groundwater Governance in South Asia* (Washington, D.C.: Resources for the Future, 2009), 310.

[6] M. Rodell, I. Velicogna, and J. S. Famiglietti, "Satellite-based estimates of groundwater depletion in India," *Nature Online* (2009), http://www.nature.com/nature/journal/v460/n7258/full/nature08238.html, accessed August 12, 2009.

[7] Shah, 2009.

[8] D. Seckler, R. Barker, and U. Amarasinghe, "Water Scarcity in the Twenty-First Century," *Water Resources Development* 15 (1999): 29–42.

[9] D. Connell, *Water Politics in the Murray-Darling Basin* (Annandale, NSW Australia: The Federation Press, 2007), 240.

[10] Ibid.

[11] Jon Gertner, "The Future is Drying Up," *The New York Times Magazine* (October 27, 2007), http://www.nytimes.com/2007/10/21/magazine/21water-t.html?_r=2, accessed September 20, 2009.

[12] H. G. Hidalgo, T. C. Piechota, and J. A. Dracup, "Alternative principal components regression procedures for dentrohydrological reconstructions," *Water Resources Research* 36 (2000): 3241–3249.

[13] Arizona Department of Water Resources RP, *Arizona Water Resources Assessment: Volume I, Inventory and Analysis* (State of Arizona, Department of Water Resources, 1994).

[14] U.S. Department of the Interior, Bureau of Reclamation, *Colorado River Interim Guidelines for Lower Basin Shortages and Coordinated Operations for Lakes Powell and Mead. U.S Bureau of Reclamation, Upper and Lower Colorado River Regions*, 2007.

[15] P. H. Gleick, P. Loh, S. V. Gomez, and J. Morrison, *California Water 2020: A Sustainable Vision* (Oakland, CA: Pacific Institute for Studies in Development Environment and Security, 1995).

[16] J. I. Morrison, S. L. Postel, and P. H. Gleick, *The Sustainable Use of Water in the Lower Colorado River Basin* (Oakland, CA: Pacific Institute for Studies in Development Environment and Security, 1996).

Chapter 3

[1] Comprehensive Assessment of Water Management in Agriculture, *Water for Food, Water for Life: A Comprehensive Assessment of Water Management in Agriculture*, D. Molden (ed.), (London: Earthscan and Colombo: International Water Management Institute, 2007).

[2] Ibid.

[3] Food and Agriculture Organization, *World Agriculture: Towards 2030/2050. Prospects for Food, Nutrition, Agriculture and Major Commodity Groups, Interim Report* (Rome: Global Perspective Studies Unit, 2006).

[4] Ibid.

[5] Comprehensive Assessment of Water Management in Agriculture, 2007.

[6] M. Falkenmark, "Peak Water—Entering an Era of Sharpening Water Shortages," *Stockholm Water Front* December (2008): 10–11.

[7] Ibid.

[8] D. Renault and W. W. Wallender, "Nutritional Water Productivity and Diets," *Agricultural Water Management* 45 (2000): 275–96.

[9] Food and Agriculture Organization, 2006.

[10] M. Kreith and C. A. Davis, *Water Inputs in California Food Production* (Sacramento, CA: Water Education Foundation, 1991).

[11] "How Biofuels Measure Up," *New Scientist* 2570 (September 23, 2006).

[12] B. C. Bates, Z. W. Kundzewicz, S. Wu, and J. P. Palutikof, Eds., *Climate Change and Water*, Technical Paper of the Intergovernmental Panel on Climate Change, IPCC Secretariat, Geneva (2008): 210.

Chapter 4

[1] G. C. Nielsen, "Agriculture and Climate Change: An Agenda for Negotiation in Copenhagen" (International Food Policy Research Institute, Focus 16, Brief 1, Washington D.C., 2009).

[2] B. C. Bates, Z. W. Kundzewicz, S. Wu, and J. P. Palutikof, Eds., *Climate Change and Water*, Technical Paper of the Intergovernmental Panel on Climate Change, IPCC Secretariat, Geneva (2008): 210.

[3] P. C. D. Milly, K. A. Dunne, and A. V. Vecchia, "Global patterns of trends in streamflow and water availability in a changing climate," *Nature* 438, 7066 (2005): 347–350.

[4] Peter Schwartz and Doug Randall, "An Abrupt Climate Change Scenario and Its Implications for United States National Security," (October 2003), http://www.accc.gv.at/pdf/pentagon_climate_change.pdf.

[5] Ibid.

[6] Iftikhar Gilani, "Himalayas not above climate change: study," *Daily Times* (June 5, 2009), http://www.dailytimes.com.pk/default.asp?page=2009\06\05\story_5-6-2009_pg7_29.

[7] UNEP, "Global outlook for ice and snow" (2009), http://www.unep.org/geo/GEO_Ice/.

[8] Peter Neiderer, Viktor Bilenko, Natasha Ershova, Hans Hurni, Sergeji Yerokhin, and Daniel Maselli, "Tracing glacier wastage in the Northern Tien Shan (Kyrgyzstan/Central Asia) over the last 40 years," *Climatic Change* 86 (2008): 227–234.

[9] W. Hagg, L. N. Braun, M. Weber, M. Becht, "Runoff modeling in glacierized Central Asian catchments for present-day and future climate," *Nordic Hydrology* (2006) 37(2): 93–105.

Chapter 5

[1] The World Bank, World Development Report: Agriculture for Development (2008), http://siteresources.worldbank.org/INTWDR2008/Resources/WDR_00_book.pdf.

[2] Martin Ravallion, Shaohua Chen, and Prem Sangraula, "New Evidence on the Urbanization of Global Poverty," Policy Research Working Paper No. 4199 (Washington: World Bank, 2007), http://econ.worldbank.org/docsearch.

[3] Comprehensive Assessment of Water Management in Agriculture, *Water for Food, Water for Life: A Comprehensive Assessment of Water Management in Agriculture*, D. Molden (ed.), (London: Earthscan and Colombo: International Water Management Institute, 2007).

[4] McKinsey and Company, "Charting our water future: economic frameworks to inform decision making," 2030 Water Resources Group, 2009.

[5] P. A. Sanchez, G. L. Denning, and G. Nziguheba, "The African Green Revolution Moves Forward," *Food Sec.* 1 (2009): 37–44.

[6] U. A. Amarasinghe, P. G. McCornick, and T. Shah, "India's Water Demand Scenarios to 2025 and 20590: A Fresh Look," (2009) 67–83. In U. A. Amarasinghe, T. Shah, and R. P. S. Malik (eds.) "Strategic Analyses of the National River Linking Project (NRLP) of India, Series 1," *India's Water Future: Scenarios and Issues* (Colombo, Sri Lanka: International Water Management Institute).

[7] Ibid.

[8] A. Mukherji, T. Facon, J. Burke, C. de Fraiture, J-M Faurès, M. Giordano, D. Molden, and T. Shah. "Revitalizing Asia's Irrigation: to sustainably meet tomorrow's food needs." Colombo, Sri Lanka: International Water Management Institute. Rome, Italy: Food and Agricultural Organization of the United Nations.

Chapter 6

[1] Malin Falkenmark, et al. "Agriculture, water and ecosystems: avoiding the costs of going too far." In David Molden, ed., *Water for Food, Water for Life: A Comprehensive Assessment of Water Management in Agriculture*. London: Earthscan and Columbo: IWMI (2007), 193–231.

[2] J. Diamond, *Collapse: How Societies Choose to Fail or Succeed* (New York: Viking, 2004).

[3] M. Leach and R. Mearns, "Poverty and Environment in Developing Countries: an Overview Study." Report to ESRC, Global Environmental Change Programme; and Overseas Development Administration. Swindon, UK: ESRC (1991).

[4] Aditi Mukherji, "The energy-irrigation nexus and its impact on groundwater markets in eastern Indo-Gangetic basin: Evidence from West Bengal," *India Energy Policy* (2007) 35(12): 6413–6430.

[5] Mark Giordano, "Global Groundwater? Issues and Solutions," *Annual Review of Environment and Resources* 34 (2009): 153–178.

[6] Agnes Quisumbing, "Male-female differences in agricultural productivity: methodological issues and empirical evidence," *World Development* (1996) 24(10): 1579–1595. Great Britain: Elsevier Science Ltd.

[7] UNEP, "Women and the Environment," (2004) www.unep.org.

[8] Caroline Sweetma, ed., *Gender and Technology*, Oxfam Focus on Gender Series (Oxford, UK: Oxfam GB, 1998).

[9] Ruth Meinzen-Dick and Margreet Zwarteveen, "Gendered participation in water management: Issues and illustrations from water users' associations in South Asia," *Agriculture and Human Values* (1998) 15: 337–345.

Chapter 7

[1] European Union, "Directive 2000/60/EC of the European Parliament and of the Council of 23 October 2000 establishing a framework for community action in the field of water policy," (Brussels, 2000).

[2] A Reuters news report on the Buriganga says it is now one of the most polluted rivers in Bangladesh, http://www.reuters.com/article/idUSTRE54I04G20090519.

[3] The Global Water Partnership was founded by the World Bank, Swedish International Agency (SIDA), and the United Nations Development Programme (UNDP) specifically as a network that promoted the coordination of water management between sectorally disparate agencies, www.gwp.org.

[4] The Global Water Partnership, www.gwp.org.

[5] World Bank, *World Development Report* (2008), www.worldbank.org/wdr2008.

[6] S. Narain, "The water-sewage connection: changing ways to the future," *5th World Water Forum* (Istanbul: Official Delegate Publication, 2009).

[7] J. Radcliffe, *Water Recycling in Australia*, Australian Academy of Technological Sciences and Engineering, Melbourne (London: Faircount Press, 2004), 232.

[8] L. Raschid-Sally and P. Jayakody, "Drivers and Characteristics of Wastewater Agriculture in Developing Countries: Results from a Global Assessment," International Water Management Institute Research Report 127, Colombo (2008).

[9] Barbara Van Koppen, "Gender and multiple-use water services, thematic note 1." In World Bank; FAO; IFAD, *Gender in Agriculture Source Book* (Washington, DC: World Bank, 2009), 235–241.

[10] Radcliffe, 2004.

Chapter 8

[1] P. Rogers and A. Hall, "Effective Water Governance," *Global Water Partnership Technical Committee (TEC)* (2003).

[2] J. Gray, "Water: a resource like any other," *Agriculture and Forestry Bulletin* (1983) 6(4): 47–49.

[3] B. Crow, "Water: Gender and Material Inequalities in the Global South," *Sustainability* (Milton Keynes, UK: Open University, 2001).

[4] J. Nehru, Jawaharlal Nehru's Speeches, vol. 2, Publications Division, Ministry of Information and Broadcasting, New Delhi (1954).

[5] E. Rap, "The Success of a Policy Model: Irrigation Management Transfer in Mexico," *Journal of Development Studies* (2006) 42(8): 1301–1324.

[6] Peter P. Mollinga and Alex Bolding, Eds., *The politics of irrigation reform: contested policy formulation and implementation in Asia, Africa and Latin America* (Ashgate, Aldershot, England, 2004).

[7] Barbara Van Koppen, "Dispossession at the interface of community-based water law and permit systems," In van Koppen, B., Giordano, M and Butterworth, J. (Eds). *Community Based Water Law and Water Resources Management Reform in Developing Countries* (CABI Publication: Wallingford, UK and Cambridge, MA, US, 2007).

[8] Douglas J. Merry, et al., "Policy and institutional reform: the art of the possible," In David Molden (ed.), *Water for Food, Water for Life: A Comprehensive Assessment of Water Management in Agriculture*. London: Earthscan and Columbo: IWMI (2007), 193–231.

Chapter 9

[1] International Business Machines, Corporation (IBM), "Water: A global innovation outlook report," (2009).

[2] Gareth Hardin, "The Tragedy of the Commons," *Science* 162 (1968): 1243–1248.

[3] The Dublin Statement on Water and Sustainable Development was released in 1992 with recommendations on action at local, national, and international levels based on four principles of water management:

Principle No. 1: Fresh water is a finite and vulnerable resource, essential to sustain life, development, and the environment.

Principle No. 2: Water development and management should be based on a participatory approach, involving users, planners, and policy-makers at all levels.

Principle No. 3: Women play a central part in the provision, management, and safeguarding of water.

Principle No. 4: Water has an economic value in all its competing uses and should be recognized as an economic good.

They have been highly influential in driving thinking on water management, particularly the idea to value the economic aspects of water and also to decentralize the management of water. In addition, they stated the importance of women to water management.

[4] Department of Water Affairs and Forestry, South African Government, "Water Supply and Sanitation Policy White Paper: Water—an indivisible national asset," 1994.

[5] H. M. MacKay, K. H. Rogers, and D. J. Roux, "Implementing the South African water policy: Holding the vision while exploring an uncharted mountain," *Water SA* 29(4) (2003), http://www.wrc.org.za.

[6] National Science and Technology Center, Bureau of Land Management. Water Appropriation Systems, 2009, http://www.blm.gov/nstc/WaterLaws/appsystems.html.

[7] Ibid.

[8] ATSE, "Water and the Australian Economy. Special Report," Australian Academy of Technological and Engineering Sciences. Melbourne, Australia (1999).

[9] *National Water Resources Audit*, National Water Commission, Canberra, Australia (2007).

[10] S. Beare and A. Heaney, "Irrigation, water quality and water rights in the Murray-Darling Basin, Australia," Australian Bureau of Agriculture and Resource Economics Conference Paper 2001/15, http://abareonlineshop.com/PdfFiles/PR11872.pdf.

[11] H. M. MacKay, 2003.

[12] T. Shah, "Issues in reforming informal water economies in low-income countries: examples from India and elsewhere," in *Community Based Water Law and Water Management Reform in Developing Countries*, ed. B.V. Koppen, M. Giordano, and J. Butterworth (Wallingford, U.K.: CAB International, 2007).

[13] McKinsey and Company, "Charting our water future: economic frameworks to inform decision making," 2030 Water Resources Group: McKinsey and Company, 2009.

[14] This paragraph is based on a personal communication with Dr. Tushaar Shah.

[15] A. Dinar, M. W. Rosegrant, and R. Meinzen-Dick, "Water allocation mechanisms," Policy Research Working Paper 1779. (Washington, DC: The World Bank, 1997).

[16] Ibid.

[17] NWC, "Water management in Australia: a vision of a positive future," 2006, http://www.nwc.gov.au/resources/documents/Water-management-Australia-vision-PUB-0906.pdf.

Chapter 10

[1] Comprehensive Assessment of Water Management in Agriculture, *Water for Food, Water for Life: A Comprehensive Assessment of Water Management in Agriculture*, D. Molden (ed.), (London: Earthscan and Colombo: International Water Management Institute, 2007).

[2] http://www.nwc.gov.au/www/html/835-new-water-sharing-data-arrangements-on-the-way.asp.

[3] Matthew Rodell, Isabella Velicogna, and James S. Famiglietti. "Satellite-based estimates of groundwater depletion in India," *Nature* 460 (2009), http://www.nature.com/nature/journal/v460/n7258/abs/nature08238.html

[4] A. Mukherji, T. Facon, J. Burke, C. de Fraiture, J-M Faurès, M. Giordano, D. Molden, and T. Shah. "Revitalizing Asia's Irrigation: to sustainably meet tomorrow's food needs." Colombo, Sri Lanka: International Water Management Institute. Rome, Italy: Food and Agricultural Organization of the United Nations.

[5] L. Raschid-Sally and P. Jayakody, "Drivers and Characteristics of Wastewater Agriculture in Developing Countries: Results from a Global Assessment," Research Report 127. (Colombo, Sri Lanka: International Water Management Institute, 2008).

[6] Ibid.

[7] WHO, "Guidelines for the safe use of wastewater, excreta and greywater," *Volume 2: Wastewater Use in Agriculture*. World Health Organization. Geneva, Switzerland, 2006. http://www.who.int/water_sanitation_health/wastewater/gsuweg2/en/index.html.

[8] P. L. Du Pisani, "Direct reclamation of potable water at Windhoek's Goreangab reclamation plant," *Desalination* (2006) 188(1–3): 79–88.

index

A

B

C

D

E

F

G

H

I

J–K–L

M

N

O–P

Q–R

S

FT Press
FINANCIAL TIMES